THE IRRATIONAL DECISION

The Irrational Decision

HOW WE GAVE COMPUTERS THE POWER TO CHOOSE FOR US

BENJAMIN RECHT

PRINCETON UNIVERSITY PRESS
PRINCETON & OXFORD

Copyright © 2026 by Princeton University Press

Princeton University Press is committed to the protection of copyright and the intellectual property our authors entrust to us. Copyright promotes the progress and integrity of knowledge created by humans. By engaging with an authorized copy of this work, you are supporting creators and the global exchange of ideas. As this work is protected by copyright, any reproduction or distribution of it in any form for any purpose requires permission; permission requests should be sent to permissions@press.princeton.edu. Ingestion or use of any Princeton University Press intellectual property for any AI purposes is strictly prohibited without a license to do so; licensing requests should be sent to digitallicense@press.princeton.edu.

Published by Princeton University Press
41 William Street, Princeton, New Jersey 08540
99 Banbury Road, Oxford OX2 6JX

press.princeton.edu

GPSR Authorized Representative: Easy Access System Europe - Mustamäe tee 50, 10621 Tallinn, Estonia, gpsr.requests@easproject.com

All Rights Reserved

ISBN 978-0-691-27244-3
ISBN (e-book) 978-0-691-27246-7

Library of Congress Control Number: 2025947003

British Library Cataloging-in-Publication Data is available

Editorial: Hallie Stebbins and Chloe Coy
Production Editorial: Jenny Wolkowicki
Jacket design: Chris Ferrante
Production: Erin Suydam
Publicity: Maria Whelan and Kate Farquhar-Thomson
Copyeditor: Annie Gottlieb

Jacket credit: Photograph © James Ball

This book has been composed in Arno Pro

Printed in the United States of America

10 9 8 7 6 5 4 3 2 1

For Cathy, who leads with her values.

CONTENTS

1 Mathematical Rationality   1

2 Searching for the Cyberphysical Utopia   20

3 This Is Not Nam. There Are Rules   64

4 Regulations, Regularities, and RCTs   102

5 When Past Performance Is Indicative of Future Results   140

6 Humans Against the Machine   182

7 Cyborg Decision Making   208

Afterword   227

Notes   231

Index   253

THE IRRATIONAL DECISION

1

Mathematical Rationality

IT TOO often feels like every choice in our daily experience, no matter how momentous, gets reduced to a risk analysis of chances and costs. Nutrition science tells us that a Mediterranean diet lowers the chance of heart disease, so we weigh the costs of heart attacks against what we like to eat. Worried parents meticulously plan the lives of children to maximize their chances of getting into good colleges. Polling tells us how pandering increases the chance a candidate wins an elected office, leading a candidate to decide how much to sell their conscience to maximize their chance of winning. Analysts now tell us that even sports can be distilled into raw numbers, with each play contributing to an accumulated probability of winning.

Is using risk management to guide every aspect of our lives putting us on a promising path toward happiness? I suppose it seems reasonable to say that there are some parts of life where risk analysis is important. Understanding the chances of financial assets gaining or losing value matters for retirement planning. You should estimate the risk of catastrophes when buying insurance. Statistical tests weighing genetics and risk factors inform prophylactic interventions in people at risk of developing serious diseases later in life. But going to restaurants? Raising children? Deciding who we love? Is every decision in life actually reducible to betting on a game of chance? Certainly not! But then why has risk management permeated every aspect of our lives, from public policy to technological acceleration to health decisions? Where does this unshakable pressure to constantly optimize come from?

Central to this idea of relentless optimization is the notion of rationality. There seems to be an overarching idea today that rationality should dictate decisions—that there are right and wrong decisions independent of the ultimate outcome. The right decision is the one that maximizes the probability of winning. Winning apparently always has a cut-and-dried definition. A rational person tabulates statistics, determines the costs of outcomes, and then bets on the option with the biggest payout. A rational person is a gambler. A rational person is an insurance agent.

If this description of rationality doesn't line up with your conception of rationality, you are not alone. Digging through scholarly works, the term is hard to pin down. Someone might get at rationality by arguing that it derives from "reason." You might say a person who behaves in a reasonable way is acting rationally. Someone else might argue that rationality just amounts to having a consistent set of beliefs. The only consensus we seem to have is that rationality is good and irrationality is probably bad. Indeed, we often end up defining rationality in terms of cultural norms around irrationality. We might think it's irrational to be afraid of dogs. We deem it irrational to gamble at casinos, as the odds are stacked against you. We all know love is irrational.

The truth is that there are and should be many ways to define rationality. But the fact is, the word "rationality" has come to mean something very specific in our modern context—almost equated with the risk-analytic mindset I've been describing. You might hear a case made like this: *Of course*, a rational person will weigh costs and benefits, estimate risks, and plan optimally. That's what it *means* to be rational. An irrational person will let emotions and personal bias influence their decisions, decisions which can easily end up being the wrong ones. If you don't want to be irrational, you'd better start doing your cost-benefit analyses, bub. Given some evidence and a set of possible actions in response to that evidence, there is always a rational way to proceed (and all other ways to proceed are thus deemed irrational). This might not lead to a win, but perhaps rationality *in and of itself* is the win.

A prime advocate for this notion of rationality is Harvard psychologist Steven Pinker. Pinker is one of the favorite intellectuals of the American elite, one of *Time* magazine's 100 Most Influential People, and frequent speaker at TED, Davos, and other hangouts of the ultra-rich and powerful. In his 2021 book, aptly entitled *Rationality*,[1] Pinker sets out to make a case for rationality and why we'd be better off if the hoi polloi was even more rational.

Pinker defines rationality as "the ability to use knowledge to attain goals." He clarifies the word knowledge to mean "justified true belief." Rather than fighting over what counts as truth,[2] let me drop the word "true" so we don't have to get into the philosophical morass of truth and meaning. I will use the following working definition: "Rationality is the ability to use justified beliefs to attain goals."

For Pinker and his powerful peers, rationality is defined with respect to goals. Rational agents must have goals. They also must be able to act. They decide to act based on whatever will most likely result in their desired goals. They decide on the most likely profitable action using their justified beliefs. Since it is so centrally about quantification, algorithms, and logic, I am going to give this conception of rationality a special name: *mathematical rationality*. Mathematical rationality captures a colloquially accepted definition that undergirds all of the rational decisions I described above. It is, in short, our modern definition of rationality, the idea that guides so many of our decisions today. Build an understanding of how diets affect your chances of a long life and pick the best one. Figure out which extracurriculars maximize students' chances of getting into college and sign your kids up for those. Mathematical rationality wins the day.

Pinker's narrow view of mathematical rationality focuses on means of decision making. We choose our actions and deploy our resources in the most "rational" way to achieve our desired ends. This view is popular among Silicon Valley executives, Wall Street traders, and a growing cadre of public intellectuals like Pinker and Nate Silver, one of the most celebrated election analysts of the twenty-first century. Silver is lauded for bringing cold, analytical data to the once vibey analysis of election horse races. Since selling his wildly popular website, FiveThirtyEight, he has pivoted to being a staunch defender of mathematical rationality as the way to understand uncertainty and succeed in life. Given how much Silver praises rationality, it should come as no surprise that he began his career as a professional gambler. In his book *On the Edge*, he writes (without evidence) that "rich and powerful people" are disproportionately the mathematically rational. The ones guided by logic, not emotion. He exclaims, "Those of us who understand the algorithms hold the trump cards."[3]

Silver's gaming analogy touches on the most recent and culturally visible influx of risk management into everything else: the phenomenon of "analytics" in sports. You might be familiar with the Brad Pitt vehicle *Moneyball*, which adapts Michael Lewis's 2003 telling of the statistical management of the Oakland Athletics to the big screen. More recently, analytics has taken over American football, where complex formulas are computed to select the appropriate plays to maximize odds. In football, a sport culturally associated with peak anti-intellectualism, many fans have bristled at this encroachment by the risk analytic mindset.

After a particularly consequential play backfired in a playoff game in January 2024, Greg Olsen, a former star player who had grown to be football's smartest color commentator, went on ESPN's *Pat McAfee Show* to champion analytics' role in football. Olsen declared:

> The outcome doesn't make the decision right or wrong. We're going to go to Vegas here in a couple of weeks for the Super Bowl. Me and Pat are going to sit up at the blackjack table and I could be the dick that sits there and hits 18 until I get a 21. If I sit there all night, everyone on the thing is going to MF me all day. They're not going to want to play with me because I'm stupid. I'm defying all logic. But one time I'm going to pull a three, and I'm going to throw a freaking party. It doesn't mean I was right. It still means I was a dumbass.
>
> The point is you could hit 16 all day. We've all been there. Just because you break 20 times doesn't mean it's the wrong decision. That's what's applying now to the game [of football] and they're trying to increase every percentage of win probability they can.[4]

There's so much to unpack in this quote. Olsen argues that there are decisions that are logically correct in sports. The logic of increasing probability is how the game needs to be played, regardless of outcome. This logic is the same as in the game of blackjack, where there is an optimal strategy for winning. Football is apparently no different from a casino game. Going against the optimal strategy is "defying all logic." It is, in his mind, *irrational*.

But the idea that there's a right way to play blackjack is actually quite new. Blackjack has history dating back to the 1600s in Europe. You might think that someone would have written down a good strategy for the game around then. Or perhaps in the 1700s. It turns out that the first attempt at coming up with an optimal strategy for blackjack was in 1956.[5] Why did it take so long?

Silver, Pinker, and Olsen's ideas about rationality—their insistence on algorithms and their references to ideal strategy—also trace back to the 1940s and 1950s, when academic scholars in mathematics, statistics, and economics formalized this narrow view of the rational. It was during this same time period that the systems we'd recognize as modern computers were first designed, built, and sold. This was no coincidence. As we will see in this book, it was precisely the design and development of computers that determined the modern definition of mathematical rationality and its application to our everyday lives. In the eyes of its very first designers, the computer would become the ideal rational agent programmed to make optimal decisions. This ideal agent is an imaginary being we've been attempting to build, with varying degrees of success, since the 1940s. What is less clear is when we decided that we should strive to live our lives as if we were computers.

Computers have advanced to the point where the statistical analyses that enable so-called optimal decisions are easier to run than ever. These analyses are applied all over the place, from logistics and planning of shipping goods to deciding which videos to offer you on streaming services. Internet platforms build their businesses on statistical analyses of their userbases, deploying advertisements and recommendations they deem most likely to increase engagement. With computing advances, weather prediction yields reliable 10-day forecasts that help people plan their lives. Statistical tests of personal risk factors inform prophylactic interventions against the development of serious disease later in life. This track record of successes might suggest to someone that it is worth emulating a computer when navigating their life.

Mathematical rationality, often called "normative decision theory" or sometimes, by economists, "rational choice theory," is the rationality of computers. It is by its very nature algorithmic, strategic, and procedural. Though I only recently started calling it mathematical rationality, I've been studying the methodological pillars of the ideal rational agent for my entire career. These pillars are the fields of mathematical statistics, optimization, control theory, and machine learning. What unites these fields is their mathematical language for computers to rationally compute optimal decisions. This language is the bedrock of mathematical rationality and formalizes the computational components of the ideal rational agent. These are huge fields, and we can't cover them comprehensively in a single book, but in the following pages we will learn what each contributes to mathematical rationality:

1. **Mathematical optimization.** A rational agent aims to act so that it achieves the best outcome given its model of the world. To do this, it writes down a mathematical description of how actions translate into outcomes. The

agent can then devise a computer program that computes the action that gives the best return under the assumptions of this model. This process of modeling and maximizing is called mathematical optimization.
2. **Game theory**. A particular case of mathematical optimization assumes that you are competing for resources against other rational actors. Methods for taking the potential behaviors of those other actors into account and strategizing against them are the foundation of game theory.
3. **Randomized experiments**. To inform mathematical models, a rational agent has to determine the impacts of actions via experimentation. In human-facing scenarios where predicting the outcome of action can be challenging, a popular approach is the randomized experiment. In a randomized experiment, the agent will engage in a variety of scenarios and choose random actions. By tabulating the outcomes of the various actions, they can then estimate which actions yield the best returns on average.
4. **Statistical prediction**. Finally, a rational agent needs to assess the probabilities of future outcomes. An agent might look at past experiences and compute the correlations between the previously observed actions and outcomes. If it uses these correlations to forecast future events, this is called statistical prediction, or, more commonly these days, *machine learning*.

All four of these pillars were formalized in the mid- to late 1940s, the brief window between the endgame of World War II and the creation of the first modern computer. The core algorithms developed in this period drive the automated decisions of our modern world, whether it be in managing supply chains, scheduling flight times, or placing advertisements on your social media feeds. Even the ideas behind impressive contemporary technologies like Artificially Intelligent chatbots were formally developed in the 1940s. This period in history produced the ideas that led directly to our current obsession with optimization, in which every life decision is posed as if it were a round at an imaginary casino, and every argument can be reduced to costs and benefits, means and ends. To understand our obsession with rationality and what to do about it, we need to go back to the beginning. We need to trace mathematical rationality back to its roots.

In this book, we will explore the four pillars of mathematical rationality in depth, from their beginning in the late 1940s to the present. These pillars were pursued by a small, connected group of characters, though each pillar has an origin story and a life of its own. Together they embody the assumption made by the rationalists of today—Nate Silver, Steven Pinker, and others—that computers can make better decisions than humans. They impact on how we view, arrange, and govern our society, aimed at improving human decisions. But evidence from the past and present will show how mathematical rationality has sweet spots. It shines when it is employed in very specific, often narrow ways. But when mathematical rationality is employed outside of that sweet spot, it bumps up against very real limits and can result in us making decisions that don't serve us well. It is imperative that we figure out what those limits are, so that we can decide when to rely on computers to make decisions and when the decision is perhaps best left to a human. To that end, as we trace the history of mathematical rationality, I'll describe when these pillars succeed and when they come up short. I'll describe how people *actually* make decisions and how automated decision making and human decision making are often in conflict. And I'll ask how humans and machines can work in tandem toward making the best decisions in the future.

This book is my attempt to answer the questions I posed at the outset of this chapter. Is mathematical rationality the right way to make decisions? Should every life decision involve distilling big data sets into simplified statistical tables? Should every decision be about profit maximization? Should mathematical rationality trump our values? Should mathematical rationality be an end in itself, justifying public policy, technological acceleration, and health decisions?

By now you will probably have guessed my answer to all of these questions: No. As the language of mathematical rationality grew more popular in the 2010s, with the rise of internet monopolies and the ubiquity of mobile computing, I increasingly realized this rational view of the world was surprisingly narrow and limited, and I wanted to understand where it came from. I wanted to explain the mathematics behind mathematical rationality. I wanted to understand how this narrow decision framework became a dominant cultural ideal. And I wanted to look to the past and future to consider possible alternative ways of approaching our lives, our governance, and our science. Writing this book prompted a foray into a critical understanding of the foundations of my field, aimed at highlighting both the sweet spots and the limits of mathematical rationality.

Cultural critic Neil Postman asserted that technology changes the meaning of our words. "It is a certainty that radical technologies create new definitions of old terms and that this process takes place without our being fully conscious of it."[6] I want to tell you how the computer changed our meaning of the word rationality.

Let me start by setting the stage for the creation of the mathematically rational agent. Our modern notions of utility, chance, and investment all began brewing in the Enlightenment, but a solidification of these ideas would come between the two world wars. A key component of this coalescence is the formalization of modern *probability*. Such formal probability would be the foundation upon which the four pillars of mathematical rationality would be built.

Mathematical probability was first invented in the seventeenth century, but it would take three hundred years to become acceptably rigorous according to the mathematicians. Initially, the mathematics of probability was conceived to analyze odds in games of chance, such as the seventeenth-century work of Blaise Pascal and Pierre de Fermat determining how to split the betting pot appropriately in dice games. In the nineteenth century, physicists and statisticians discovered probability could be used for more than just gambling; it could help people to draw connections between the microscopic and the macroscopic.

For example, in the development of thermodynamics, physicists modeled the microscopic world as a bunch of tiny particles bouncing around. The motions of these trillions of trillions of microscopic particles appeared random. However, from far away, the blur of these tiny random particles looked *deterministic*. Physicists realized that their laboratory measurements were a blurry view of complex random activity. That is, orderly macroscopic measurements were *averages* of disordered microscopic activity.

It was during this time that we realized the world is unpredictable from certain vantage points yet highly predictable from others. For instance, I can't tell you whether a particular coin flip will land heads or tails, but I can tell you that if I flip enough times, I should see approximately the same number of heads and tails. I can't tell you how the individual particles of gas move in a balloon, but I can tell you how big the balloon will be if I put in a certain amount of helium. I can't precisely predict the static in a radio transmission, but I can tell you roughly how loud it will be at different times of day due to atmospheric interference. This is the magic of probabilistic thinking: we trade in particulars for expectations. I can't predict specific outcomes, but I can predict general behavior.

These ideas were brewing in the physics community in the nineteenth century, and statisticians contemporaneously applied them to understand variations in plant, animal, and human populations. Individuals were random and wildly varied, but the averages of the population into statistical factoids were orderly and predictable. State bureaucrats could tabulate statistics to tell us stable facts about society, whether they be about crime, mortality, or demographics.[7] Through statistics, chaos was aggregated into order.

Though these seminal ideas percolated for almost 300 years in various forms, it wasn't until the 1930s that we packed all of these related threads into a universal mathematical language of uncertainty. Russian mathematician Andrey Kolmogorov was the first to make probability into what mathematicians called a "rigorous theory" in 1933. Kolmogorov's axiomatization inspired a revolution in mathematics. For the first time, we could understand physical processes as *random* processes, ones whose measurements would always be unpredictable at the finest scales but whose behavior could be understood at a higher level of abstraction. If our interactions with the world were random yet predictable, maybe all uncertainty could be put on the same rigorous footing. With a rigorous calculus for the uncertain, statisticians concluded they could reason about *all* uncertainty in the world. They could quantify uncertainty in measurements like censuses and polls. They could understand complex random processes to remove noise in communication systems and understand properties of the subatomic. There was an emerging promise of control: If we could understand uncertainty, perhaps we could use it for our purposes. Randomness itself could be *useful*. You could use random sampling to simplify quality control, looking only at a small, representative set of products on the assembly line. You could use random tests to see if one fertilizer was better than any other. You could use randomness to deceive opponents by choosing your next action at random. Statisticians began to develop means to inform decisions despite the formidable uncertainty of existence.

As the 1930s drew to a close, it seemed we were finally at a time where we could achieve the promise of Enlightenment science. We could count things, establish statistics, draw rigorous inferences, and understand our natural world through a language of mathematics. And that mathematics could inform how to make the most critical decisions even in the face of uncertainty. We could characterize all the properties of the unknown we'd need to make appropriate, informed decisions. In the 1940s, this mathematical harnessing of uncertainty would be put to the test in one of humanity's largest, deadliest conflicts.

World War II reshaped global society, discontinuously reconfiguring borders, ideologies, and technologies. The war also transformed the relationship of natural sciences and government, drafting the efforts of pure science into applied military efforts. The United States established the National Defense Research Committee in 1941. The NDRC was the brainchild of Vannevar

Bush, an electrical engineer who was MIT's Dean of Engineering in the 1930s. Bush determined hard science and engineering was necessary to advance defense logistics against the new sophisticated technology being developed by Nazi Germany. Through Bush's academic connections, World War II saw an unprecedented involvement of civilian scientists, engineers, mathematicians, and economists. Bush argued that we should gather the theoretical and technological advances of the first half of the twentieth century and make them actionable in reality to fight the greatest threat of the second half of the twentieth century. Military branches themselves established units for tabulation, statistics, and planning. The fevered war machinations tapped technological and intellectual innovations from the prior decades, advancing them for directed military applications.

Mathematicians brought two major innovations to the war effort. The first was the hardware to run large-scale computations. In the early 1930s, Bush had built a nascent computer called a "differential analyzer." This machine could add and subtract and showed the promise of automating tedious calculations to solve complex applied-mathematics problems. One of Bush's graduate students found the differential analyzer to be too temperamental because of its analog design. In his master's thesis, he proposed an alternative means of building computers with *digital* logic gates. This master's thesis would form the basis of all of our modern computer chip architectures. Its author, Claude Shannon, would become the most influential electrical engineer of all time, single-handedly creating half a dozen engineering fields.

Inspired by these early computing successes, Bush lobbied the government for massive funds for the development of digital devices to accelerate military-driven calculations. Intricate, complex computations were done by teams of people feeding newly developed electronic calculators with punch cards, supervised by top mathematicians. These calculation efforts helped predict the behavior of complex rocket ballistics and antiaircraft weaponry and were critical to the design of the atomic bomb. Though it wouldn't be finished until after the war ended, the ENIAC, what many consider the first programmable computer, was commissioned during the war to further accelerate such calculations.[8]

A parallel contribution of mathematics to the war effort was the application of probability and statistics for logistics and planning. There were innovations in how to communicate securely and break codes. New techniques were developed to predict the behavior of enemy aircraft and determine what sorts of ships were present in noisy radar signals. Because of advances in therapies, militaries began recording the results of large-scale field trials in managing disease, infection, and wounds.

At the center of all this activity was a group of mathematicians who would become famous in the aftermath of the war. The most famous of them, Norbert Wiener, John von Neumann, and Claude Shannon, will be central to the story of this book. Before World War II, Wiener and von Neumann were well known among mathematicians for their contributions to the foundations of probability. Wiener was one of the founders of the theory of stochastic processes, and von Neumann did seminal work in quantum mechanics and what would become game theory. The war experience opened their eyes to what the abstract frameworks they built could do for concrete reality. Mathematicians saw firsthand how quantified predictions of the unknown could improve our decision-making capabilities.

Mathematicians left the war inspired by their ability to have practical impact. The world needed rebuilding, and they wanted to be part of that rebuilding effort. They aimed to take the ideas that had shown promise in the war and apply them toward peacetime efforts. Toward defense. Toward public health. Toward governance.

The key mathematical insight of the war was linking that which could be calculated to that which could be automated. Mathematicians were tantalized by the possibility that if humans could make a decision, then that decision not only could also be made by machines, but could possibly be better made by machines. Uncertainty would still exist, of course, but quantification of uncertainty could be made methodical and rule-based. If decisions could be calculated from such quantified uncertainty, then they too could be automated. These decisions would be more "scientific." More "rational." The mathematicians of the day set out to build rational machines.

Inspired by the success of mathematical logistics in the war effort, this small group of talented mathematicians dedicated the decade after the war to fully automating mathematically rational decision making. This would require not only new techniques, but also new machines that did this automation. There was thus a rapid co-development of the abstraction of rational decision making and the design and construction of the architectures and circuits of the computer.

As we will see in this book, the codified mathematics of the 1930s that had proven so valuable in the war inspired the codification of an ideal rational agent. To make decisions in the face of adversity and uncertainty, uncertainty needed to be quantified. The ideal rational agent would do just that by equating the unknown with the random. Just as nineteenth-century physicists had shown, a vast collection of unpredictable events could look predictable on average. Once uncertainty was quantified, a plan would need to be constructed to maximize the chances of a good outcome. The value of outcomes would be equated with a universal currency. A rational agent would thus assess the probability of various futures and choose its action to maximize its returns. It would be less ad hoc, temperamental, and tied to conventions than its flawed human designers. It would be mathematically rational.

Between 1945 and 1950, not only was this mathematical ideology fully developed, but the modern computer architecture was designed and built to execute it. Computers were designed to build better weapons and make better decisions in the shift from the World War to the Cold War. The rise of the Cold War administrative state, with massive investment in research, academia, and medicine, provided the ample funds and resources needed to build these first computers. And the computers would be designed to execute the four pillars of mathematical rationality: mathematical optimization, game theoretic strategy, randomized experimentation, and statistical prediction.

And though our contemporary computers are billions of times more dense and powerful than their 1950s counterparts, their internal logic, organization, architecture, and robotic ideology have remained the same.

The centering of the computer and its theoretical potential explains how mathematical rationality shaped how we conceive of rationality today. Part of the power of mathematical rationality is attributable to the unfathomable growth of computing. For a long time, computers got twice as fast every two years.

Gordon Moore, the co-founder of Intel, predicted this in 1965, and for decades the computing industry proved Moore correct. Such exponential growth meant that problems that seemed unsolvable last year became solvable the next year. From 1965 to 2015, fifty years after Moore's prophecy, computing power doubled 25 times, equal to a mind-boggling factor of over 30 million. Computers got faster, and that gave us a guiding vision. We could see how far the information age could take us.

Growth in computer "intelligence" far outgrew growth in human intelligence. And since there was so much excitement about these improved business engines, it didn't really make sense to worry too much about alternative ways to make decisions. Adding more computing and data seemed to enhance decision making at multiple points. Computers helped governments grow and manage a blooming administrative state. Computer backends help manage businesses. Computers created a globally connected financial system that has enabled unprecedentedly interconnected trading and commerce. They undergird path dependencies in our complex and confusing healthcare system. There is nothing we do anymore that doesn't touch a computer. The logic of spreadsheets, the immediacy of email, and the information of the internet all shape how we think about knowledge. We equated the booming power of our computer systems with the power of mathematical rationality to solve problems.

But if rational decision making is a hammer, every decision looks like a nail. There are sweet spots for each pillar of mathematical rationality, but it's easy to get trapped by one's tools and fail to realize that the most difficult questions are the ones the tools can't answer. Mathematical rationality is peculiar and robotic! All unforeseen occurrences are deemed conceptually equivalent to a lottery. All decision making is a proper mechanical analysis of risk. The mathematically rational agent always identifies and chooses the least risky potential outcome. According to the inventors of computers, the truly rational agent is an actuary.

Today, the unprecedented scaling phase of the information age is ending. Access to the computing power needed to solve the most daunting contemporary calculations is concentrated among a few large tech companies. And even for these folks, exponential scaling must end too. We'll run out of data centers and electricity sooner rather than later. So what do we do next? Maybe it's time to step back and ask whether we're happy with the shortcomings of our big bureaucratic system. To ask why we're stuck with our idiosyncratic mathematically rational decision-making paradigm. To look to other ways of making decisions.

Steven Pinker and Nate Silver both claim we should strive for more mathematical rationality. I will present the case for why the future really needs less. We should use the tools of mathematical rationality only sometimes, in the sweet spot where they do make very good decisions—but the rest of the time, we need a human touch. Sure, humans aren't mathematically rational, but some decisions can't be reduced to numerical calculations. Sometimes our ingenious human qualitative irrationality is precisely what a situation needs. In the pages that follow, I'm going to show you how the four pillars of mathematical rationality came to such power and ubiquitousness—but also how they can only get us so far.

2

Searching for the Cyberphysical Utopia

EARLY IN America's engagement in World War II, the Army Air Forces stood up a data science unit. The Statistical Control Division did everything we expect a good data science team to do: "collecting, processing, analyzing, and presenting statistics" of capabilities and activities.[1] This data collection was aimed at providing quantitative insights into the logistics and planning of the war effort, including where bombers should be deployed and what equipment needed to be manufactured. For example, the division's careful tabulation saved the Air Force billions of dollars after finding they were producing more bombs than they could possibly drop.

George Dantzig was a civilian analyst at the Statistical Control Division, running the Combat Analysis Branch.[2] Dantzig was tasked with understanding how the availability of varied resources in the war, be they people, munitions, or fuel, contributed to successes and failures in combat. Dantzig developed detailed surveys to estimate these variables and hoped to find plans that would result in the optimal balance, providing the right resources to the right places to guarantee American success in air operations. Dantzig ran into many frustrating roadblocks when trying to build systematic plans from the mountains of tabulated data. First, completing plans took so long that they would be obsolete before they were even finished. Plans would often be roadblocked, with many possible alternatives and no clear way to choose between them. In these cases, decisions would be ad hoc but given authority by calling them "mature judgments."[3] Dantzig was even more frustrated by the muddled nature of planning. As he would later put it:

> Ask a military commander what the goal is and he will say "The goal is to win the war." Upon being pressed to be more explicit, a Navy man will say "The way to win the war is to build battleships," or, if he is an Air Force general, he will say "The way to win is to build a great fleet of bombers."[4]

In Dantzig's experience, military officers often forgot about higher-level objectives, and the resources under their purview became the goals.

Dissatisfied, Dantzig wanted a framework to systematize and automate the Air Force's decision making in what he described as "a rapidly changing economy."[5] He thought he should be able to use his surveys, which measured the inflow and outflow of available resources, to mathematically derive the best course of action for the next phase of operations. He called this problem of mathematically deriving optimal activities from data about resources and constraints *programming*. Dantzig's programming formalized our modern mathematical theory of optimization.

In our contemporary world obsessed with optimizing everything, it is hard to appreciate how radical the conceptualization was in the 1940s. The idea that a person could automatically calculate an optimal plan simply by being disciplined about how they wrote down their knowledge transformed how we thought about the world and would revolutionize postwar policymaking.

Dantzig's notion of programming has little to do with computer programming as we know it today. Dantzig's programming declares what we *can* do and what we *want* to do. A computer program, by contrast, tells a machine *how* to do a job, listing instructions for a computer to execute in sequence. But in his quest to build machines to automate optimal programs, Dantzig would personally influence—both directly and indirectly—the design of our primary means of automation, the computer itself. The development of optimization gave rise to the computer, and computers then gave rise to new forms of optimization. The two would complement each in a whirlwind feedback loop, leading to decades of exponential growth.

Mathematical optimization is now employed to manage supply chains, route airlines, control robots, and train neural networks. Few industries are untouched by Dantzig's formalization of programming. This chapter will trace the roots of this field that is today critical to how our society functions, demonstrating how it was inextricably intertwined with the creation and scaling of the modern computer. Our discussion will take us through seventy-five years of the history of this field, starting with Dantzig and ending with modern microprocessor design. We will see how the desire to optimize required more advanced programming techniques, which then required more advanced computer hardware. In a similar vein, more advanced programming techniques let us design more advanced computers. This feedback loop would flourish for decades, ending in computing infrastructure beyond anything Dantzig could have imagined. Ultimately, we'll end in the present, where we'll appreciate both the power and limits of optimization. Optimization in its sweet spot ushered in some of the most impressive advances of the optimization age. Optimization when misapplied often led to catastrophe. *When is safe to optimize* is the central question in a world dominated by technology.

To set the stage for Dantzig's breakthrough, let's first walk through how to formulate an optimization problem. Considering the reality of some *situation*, how do we maximize an *outcome* by taking an appropriate *action*? Surmising a situation and declaring an action is the prescription of a *policy*. Finding the policy that maximizes outcomes is the goal of optimization.

Think about a problem we all face at the grocery store: What is the most budget-efficient diet? You want to maximize your bank account balance at the end of the week, but you also want to make sure you get enough protein and essential minerals like iron and potassium, without too much salt or unhealthy fats. What is the best set of groceries to buy?

This grocery budgeting example illustrates the component parts of an optimization problem. There is a *goal*, namely maximizing the balance in your bank account. There are *constraints*, namely, all of your dietary needs. How can we translate these goals and constraints into an optimal policy that maximizes the goal while satisfying all the constraints?

To pose a goal mathematically, we need to have some formula for computing the value of any particular policy. In the grocery budgeting problem, the value is an amount of money, but any abstract notion of value would work. The constraints need to be formulated as mathematical expressions that are true when the policy satisfies the constraints and false otherwise. Goals and constraints are colloquial terms that we all can conceptualize, but how can we know when goals are mathematically modelable? And even if we can formulate our goals as math expressions, can we find the best policies systematically? Can we guarantee that we have found the best policy out there?

Dantzig proposed the first broadly applicable optimization formulation that answered all three of these questions affirmatively. Dantzig proposed *linear programming* as the simplest "nontrivial" mathematical optimization problem. It is *simple* because the goals and constraints are *linear*, meaning they are special mathematical formulae that can only model certain phenomena. Linear programming is *nontrivial* because it is only practically solvable with the assistance of a computer. However, with a computer, not only can you find the optimal policy, but you can find a certification that this policy is optimal and no other policy out there can do better.

The grocery budgeting problem turns out to be an example of a linear program. To turn the problem into a mathematical model, take all the foods you'll eat this week. The total cost is the sum of the individual costs on your grocery bill. Each item on that bill is the quantity of the item times its cost. This cost is *linear* with respect to the amount of each food in your diet: If we fixed all of the items but one and made a plot of the cost as we varied the amount of this one item, we'd generate a line.

Similarly, the nutrient constraints are linear. The amount of each nutrient is the sum of the nutrient contributions of all the foods in your diet, and the contribution of each individual food is the amount of food times the amount of nutrients per unit of that food. For example, an egg has 72 calories and 6 grams of protein. If you eat three eggs, you are $3 \times 72 = 216$ calories on your way to the total calories for the day and 18 grams toward your protein goal. We could formulate similar constraints for vitamins and minerals, and these would be linear as well.

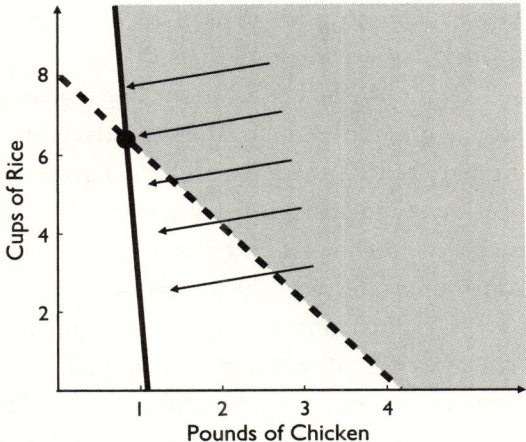

FIGURE 1. Optimizing a simple diet by linear programming.

Having modeled the budgeting problem as a linear cost to be maximized and a collection of linear constraints to be satisfied, how do you find the most cost-effective diet?

Dantzig's crucial discovery was the second step: a general-purpose procedure for finding the optimal solution of any linear program. Dantzig called his algorithm "the simplex method." This is an unfortunate name, and one Dantzig almost dismisses on the last page of his report on the method.[6] A better name would be "the basis method."[7] A basis is a set of policies that can be mixed together to create a variety of other policies. Every policy in a linear program can be written as a combination of bases.

The policies of a diet planning problem can be graphed to illustrate the role of bases. An oversimplified version of the diet problem is depicted here in figure 1. Let's just look at two foods, chicken and rice, and two nutritional goals, calories, and protein. The horizontal axis is the pounds of chicken consumed per day and the vertical axis is the cups of cooked rice. The region above the thick dashed line is all diets that have at least 2,000 calories. The region above the thick solid line is all the diets that have at least 100 grams of protein. The shaded region is all of the diets that satisfy both dietary constraints, having at least this amount of calories and protein. The arrows denote the direction of decreasing cost. The lowest-cost diet occurs at a corner where the two constraints intersect. This is precisely one of Dantzig's bases. And if you added more foods and more constraints, you'd end up with a myriad collection of such bases.

Each step of the simplex method takes a candidate basis in hopes that the optimal policy can be found by combining the elements of the basis. If it can't find the optimal policy, it removes one policy from the basis, adds in another, and tries again. Dantzig showed that this iterative procedure of trying out bases will eventually find the optimal policy.

Figure 2 illustrates the dynamics of the simplex method. In this optimization problem, there are three constraints on the variables depicted by dashed lines. A solution is feasible if it lies below all three lines. The set of all feasible solutions is the shaded region. There are five bases, denoted A, B, C, D, E. Let's suppose the simplex method starts at basis A. It then checks for a direction of improvement, either moving to basis E or B. Since B has a higher return it moves to vertex B. B has adjacent vertices A and C. Only C has higher return and the method proceeds to C. Finally, vertex C has two adjacent vertices, B and D. Neither has higher return and the simplex method terminates. Though this picture is drawn in two dimensions, the same procedure works in higher dimensions, hopping vertex to vertex to increase the reward until no improvement is possible. Dantzig's critical insight was that these updates could all be done numerically by solving systems of linear equations.

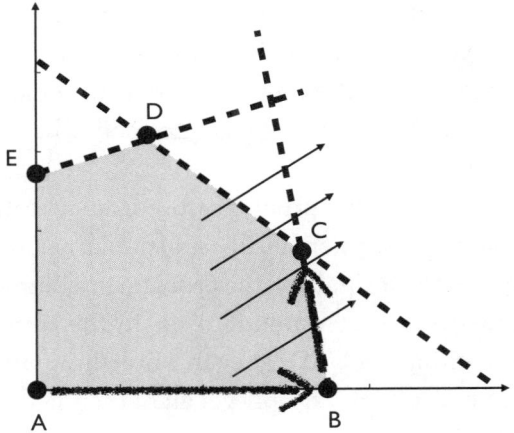

FIGURE 2. The dynamics of the simplex method.

If a modeler can pose their policy optimization program as a linear goal with linear constraints, Dantzig's simplex method will find the answer. Dantzig's procedure is easy to write down in mathematical expressions, taking only a page or so in his original publication. But computing the numerical value of these expressions required intense calculations, and the intractability of computing these solutions by hand was an early driving force behind the development of computers.

After deriving the simplex method, Dantzig was keen to find a candidate benchmark linear program to test. After consultation with various analysts at the Bureau of Labor Statistics and the Department of Agriculture, Dantzig settled on the problem I just used to illustrate mathematical optimization: the grocery budgeting problem! Optimizers today know this problem as "the minimum cost diet problem." The diet problem is still one of the first linear programs taught in optimization classes. Less heralded is the story of the origins of the diet problem, which remains an illustrative way to explain not only what optimization can do but what it can't.

The minimum cost diet problem arose from a heated battle about the role of government in advising public health. Economist George Stigler[8] formulated the problem in a biting critique of USDA nutrition recommendations. In the midst of the Great Depression, the USDA began publishing nutritional guidelines to help people cheaply get all of their necessary nutrients. The major vitamins and nutrients had been established by the 1930s, and economist Hazel Stiebeling worked to find affordable diets for most people that contained all of the recommended allotments of calcium, iron, and vitamins.

Stiebeling and her co-author Rowena Carpenter described a minimum cost recommendation as "the cheapest combination of foods that it is desirable for an indefinite period."[9] Stigler took issue with such claims, complaining that Stiebeling and her colleagues weren't being scientific: For instance, how do we know what is "desirable"? He set out to decouple "nutrient subsistence needs" from "non-scientific" concepts such as how food tastes. To do this, he formulated a mathematical optimization problem to find the diet that was *actually* of minimal cost.

Stigler emphasized that the diet problem is a gross oversimplification of the role of food in nutrition. The recommended daily allowances of nutrients set by the National Research Council were mostly guesses, and many of these guesses were inflated by 50 percent just to be safe. The extent of individual variation in nutritional needs was unknown. The amount of nutrients in cooked food would vary based on the preparation. Even different varieties of apples vary wildly in their nutritional content. The costs of foods vary by season. Stigler worried there was an "almost infinite complexity of a refined and accurate assessment of nutritive value of a diet."

Setting his major caveats aside, Stigler assembled a list of foods and their 1939 prices from the Bureau of Labor Statistics. He gathered the nutrient content from the Department of Agriculture. He compiled a list of 77 possible foods and had 9 nutritional requirements for calories, protein, calcium, iron, and vitamins A, B_1, B_2, B_3, and C. Stigler then set out to find the minimum-cost diet.

Finding this diet was a challenge. Stigler had pieced together a linear program without knowing what linear programming was. Dantzig's simplex method would not be available for a few more years, and Stigler would have to rely on intuition and heuristics. At first blush, Stigler worried he might need to consider every possible combination of the foods. For example, he used common sense not to add white bread to the diet, as one could add wheat flour and bake bread. Still, he lamented that his "procedure [was] experimental because there does not appear to be any direct method of finding the minimum of a linear function subject to linear conditions." He winnowed the list of foods to 9: wheat flour, evaporated milk, cheddar cheese, beef liver, cabbage, spinach, sweet potatoes, dried lima and navy beans. In Dantzig's terms, he had found a basis for the optimal policy by introspection alone. Stigler then tried several combinations by hand and found a diet that cost $39.93 per year (which would be about $900 today). By contrast, Stigler calculated that Stiebeling and Carpenter's minimum-cost diet would run about $100. His diet was nearly three times cheaper.

Having discussed Stigler's paper with other government officials, Dantzig was convinced he could find an even cheaper diet if he applied his simplex method to Stigler's diet problem.

In consultation with Dantzig in 1948, the Mathematical Tables Project solved the linear program by hand. Based at the National Bureau of Standards (abbreviated NBS and now known as NIST) and started by the Works Progress Administration during the Great Depression, the Mathematical Tables Project was a team of hundreds of people who computed immense tables of mathematical integrals, esoteric constants, and combinatorics, with the hope that they could aid scientists and engineers. The Mathematical Tables Project combined hand-fed calculators with human organization and was ideally equipped to try to coordinate and run Dantzig's algorithm.

The Mathematical Tables Project took 120 person-days to solve the diet problem. Their solution involved gluing sheets of paper together to ensure the full calculation was legible. According to Dantzig, this tablecloth remained displayed at the NBS for many years. After dedicated calculations, the team proved that the optimal diet would cost only $39.69.[10] From a practical perspective, it wasn't worth the 120 hours to save a quarter. But it was worth the effort as proof of principle. Solving the linear program revealed that you didn't need a Nobel Prize–winning economist manually tinkering with the data to find a decent solution to complicated policy problems. The algorithm could replace and automate certain forms of mathematical expertise and cleverness. Planners could focus their expertise on the *formulation* of policy problems rather than on their *solution*. This evidence was enough for Dantzig to argue for more sophisticated hardware to solve linear programs.

Before continuing with Dantzig's quest for computers, it's worth first pausing to ask what was in Stigler's diet in the first place. In 1939, $40 a year guaranteed a daily regimen of one pound of flour, two ounces of evaporated milk, five ounces of cabbage, one ounce of spinach, and five cans of navy beans. This is even more preposterous than my imaginary illustration diet consisting only of chicken and rice. While there could be a few tasty recipes you could throw together with this diet,[11] perpetually eating five cans of navy beans a day would be miserable. The solution found by solving the linear program with the simplex method largely contained the same foods as Stigler's manual solution, only replacing the weekly evaporated milk with a morsel of beef liver each month. Dantzig's method guarantees a basis solution—simple, and elegant, and satisfying all the constraints. It does not guarantee that the solution is actionable in practice when moving from the mathematical model to the actual application.

The diet proposed by Stiebeling and Carpenter, the female economists derided by Stigler, cost about a hundred dollars a year. By the cold mathematical calculations of Stigler, Dantzig, and the Mathematical Tables Project, it was certainly not "minimum cost." On the other hand, Stiebeling and Carpenter's diet included fruits, meat, and sugar. They included $15 per year in the budget for accessories like coffee, tea, baking powder, and spices. Their diet was designed to satisfy the needs of the soul as well as the needs of nutrition. But Stigler objected:

> These cultural judgments, while they appear modest enough to government employees and even to college professors, can never be valid in such a general form. No one can now say with any certainty what the cultural requirements of a particular person may be, and on its face it will always be impossible to determine a unique cultural minimum diet for 140 million Americans of transcendental variety of background, social position, and cultural values. If the dieticians persist in presenting minimum diets, they should at least report separately the physical and cultural components of these diets.

Stigler is right, of course. Cultural norms are associated with diet, and the USDA was undoubtedly aware of this. But Stiebeling wanted to educate people about how to think about nutritional requirements while still maintaining the cultural centerpiece of family meals and breaking bread with friends. She meant "minimum cost" colloquially, not literally. The horror of actualizing minimum cost, directly optimizing the myopic objective of dollars spent, is reflected in Stigler's austere rations.

Stigler and Stiebeling's disagreement about dietary recommendations would be reflected in their future paths. Stiebeling would go on to be a leader in nutrition, serving as the Director of the Institute of Home Economics and the Deputy Administrator of the Agricultural Research Service at the USDA. She helped develop the first school lunch programs and popularized increased consumption of fruits and vegetables. In a footnote, Stigler called her a tax-supported bureaucrat beholden to government interests.[12] A graduate of the University of Chicago, Stigler would remain deeply critical of the role of special interests in government, formulating the concept of regulatory capture and eventually winning the Nobel Prize in economics.

Stigler's diet problem was the first linear program ever solved and remains the first taught in most optimization classes. But we seldom complicate its role in optimization. The nutrition modeling behind the diet problem is, at best, an approximation. The prices can vary wildly—even week to week—and if everyone decided to buy navy beans, their price would definitely go up. In short, the diet problem is a perfect example of where narrowly focused optimization can lead you. No one would eat Stigler's diet unless under extreme duress, and no modification would make this diet more attractive without significantly increasing its price. There is no single mathematical formula we can optimize to find the perfect diet.

Even the simplest optimization model we present in our optimization classes is loaded with approximations, idealizations, and value judgments. Perhaps unsurprisingly, these issues do not go away when we try to optimize more complex and challenging problems. Optimization problems are all based on idealized models of reality, and objectives are seldom easy to specify as a single mathematical formula. We have to interpret the policy suggestions produced by optimization solvers accordingly.

Dantzig first presented his formulation and solution of linear programming at a meeting of the Econometric Society at the University of Wisconsin in Madison in 1948. Dozens of economic luminaries attended, and the potential of the simplex method for economic and policy problems enthralled the conference. Dantzig had uncovered a path toward practical maximization of utility, and economists were keen to find ways to apply his ideas.

However, not everyone was happy with Dantzig's talk. The economist Harold Hotelling objected, "We know the world is nonlinear." Hotelling would be the first, but not the last, to raise such objections about optimization. Of course, Hotelling was right. The technical definition of nonlinear is "not linear." Our irreducibly complex world is undoubtedly not well described by simple linear relationships like those in the minimum cost diet problem. There is always a sharp divide between the problems we can solve and those that accurately address the complexity of the real world.

John von Neumann, also in the audience of Dantzig's talk, recognized, however, that simple models can be exceptionally useful when they are applicable. Von Neumann impatiently responded to Hotelling before Dantzig could answer: "The speaker titled his talk 'Linear Programming.' Then he carefully stated his axioms. If you have an application that satisfies the axioms, use it. If it does not, then don't."[13]

Linear programming initiated a cycle in computing where technological and conceptual constraints would create a form of selection bias in problem-solving. Such cycles begin when researchers find a novel application where computers excel. Though this application can't solve all our problems, engineers immediately try to adapt it to solve as many of their problems as possible. And why not try to take advantage of our most cutting-edge technologies? If that means leaving some desiderata aside to work on in the future, so be it. Phrased differently, all algorithmic platforms come with a set of "axioms" where they are guaranteed (or at least rather likely) to work. There is nothing stopping engineers from constraining their system to always be in a state where these axioms hold true.

In the case of linear programming, which demands linear costs and linear constraints, policy designers must guarantee that their system always respects linear constraints if they want to use the simplex method. This might mean building safeguards that ensure linearity assumptions aren't violated. In other words, the optimization axioms become a constraint on *system design*. So while the world is generally *not* linear, if a policymaker can envision a set of rules and principles where system operation *would be* linear, they can find optimal policies for their systems under these rigid constraints. Put simply, once engineers find regimes where we can design optimal policies, they design their systems to stay in these regimes. As we will see in the case of computer chips, this sort of design can lead to incredible innovations.

A hidden upside of working within small axiomatic frameworks is the benefit this constraint bestows on creativity. Limiting the set of tools at an engineer's disposal can often paradoxically make it easier to design systems. Having too many choices can lead to far too many options to evaluate and can increase the difficulty of finding a good policy. A constrained toolkit forces a policymaker to be more creative with the options at their disposal. Design constraints can inspire engineers to innovate.

The final line of the abstract of Dantzig's talk previewed what would become another cliché of computing. "It is proposed that computational techniques such as those developed by von Neumann and the author be used in connection with large-scale digital computers to implement the solution of programming problems." Though we associate Big Data with the twenty-first century, we've been obsessed with "large scale" since the dawn of computing. Computers were always designed to solve problems currently out of reach of current technological means. From the beginning, large scale has meant whatever a programmer can't easily solve with their currently available tools. This quest for the slightly out-of-reach—and the promise that solving the out-of-reach will lead to better outcomes—has driven computing forward since its inception.

Invigorated by the excitement of the Madison workshop, Dantzig was determined to find computing hardware that would solve linear programs. The hand-crafted solution of the Mathematical Tables Project would not suffice. In his role at the Air Force, Dantzig lobbied hard for financing the development of computing systems. He would later reflect:

This outpouring between the years of 1947–1950 coincided with the first building of digital computers. The computer became the tool that made the application of linear programming possible. Everywhere we looked, we found practical applications that no one earlier could have posed seriously as optimization problems because solving them by hand computation would have been out of the question. By good luck, clever algorithms in conjunction with computer development gave early promise that linear programming would become a practical science.

In 1949, with Michael Montalbano of the NBS, Dantzig demonstrated the potential of a machine that could automate the solution of linear programs. The computational setup was elaborate. It used two IBM mechanical punch calculators manually wired to perform strings of over 50 highly specific computations. They arranged these calculators in a circle of machines incorporating a menagerie of punch card sorters, reproducers, tabulators, and collators. Running their jury-rigged computational device for eight hours straight, they solved a small linear program associated with war planning.[14]

Dantzig used this demonstration to argue for a machine that wouldn't require such heroic custom engineering. Dantzig had been in frequent conversations with John von Neumann and was convinced that von Neumann's "stored program" architecture was critical for the simplex method. This architecture entailed preloading all of the data onto magnetic memory and enabling the computer to manipulate and change the memory as it ran. In theory, it would allow for the solution of much larger problems.

In hopes of producing such a machine, Dantzig continued his collaborations with the National Bureau of Standards, raising over half a million dollars of support from the Air Force and offering detailed suggestions on how their in-house computing could be designed to solve linear programs. The resulting computing system, the SEAC, debuted at the NBS in 1950. The SEAC was the first functional stored-program computer and is the prototypical architecture of the computer we know today.[15] Though intended only to be a prototype, it was an influential system. Computational image processing, weather forecasting, and chemistry were all first demonstrated on the SEAC. In 1951, the SEAC solved a linear program with 100 variables and 19 constraints in merely three minutes.[16]

Though dramatically faster than what had been possible by previous efforts, this hundred-variable problem pushed up against the limits of the size of problems the SEAC could handle. Running larger problems would require more capable computers.[17] The constant frustration with computer capabilities is a familiar one to programmers. The constraints of the architecture of the day always dictate the scale and shape of what computer programmers solve. How much memory do I have? How fast is my data transfer? How many hours would it take to solve the problem? There is always pressure to solve a problem slightly more complex than what the current computer can handle. This pressure fueled the booming computer industry for seventy years. In particular, with the potential applications rapidly growing, a company with a decades-long legacy building machines to calculate and tabulate jumped in to produce machines that could compute.

Before becoming the epicenter of high-performance computer manufacturing in the United States, IBM had built business calculators for decades. They had deep-rooted connections to the military, further cultivated during the Second World War. So when the military hoped to invest in computing technology for their Cold War efforts, IBM's leap from massive, hand-fed calculators to a stored-memory computer wasn't large. In 1951, they shipped their first computer mainframe.

In 1954, IBM unveiled their first scientific supercomputer, the 704, adding a *floating point unit* to assist with mathematical calculations. Floating point represents numbers to facilitate mathematical operations like addition, multiplication, and division. It is the convention we still most commonly use to represent decimal numbers as a string of binary bits. With this representation, it became possible to solve *linear algebra* problems on computers. Linear algebra is the manipulation of arrays of data to solve problems like linear equations, those needed to solve Dantzig's simplex method. Linear algebra remains the primary backbone of scientific computing, and it is also a fundamental tool in processing complex data such as images, sound, and even text. The inclusion of floating point capabilities in the 704 thus enabled a variety of powerful algorithms to be tried for the first time.

The technology first attempted on the 704 was mind-blowing. We'll touch on these technologies in more depth in later chapters, but for now, let's establish a timeline for these early innovations made on the 704. In 1956, the first video game, a program that played checkers, was demoed. A competent chess program would be tested in 1958. The first modern artificial intelligence program, the Perceptron, was also demonstrated on the 704 in 1958. The 704 was used in the first optical character recognition experiments at Bell Labs in 1959. The first computer music synthesizers were built on the 704, and the first demonstration of voice synthesis was on the IBM 7094, a successor to the 704. The 7094 would sing the song "Daisy Bell," and be later featured in Stanley Kubrick's *2001: A Space Odyssey*.

But IBM did not add state-of-the-art math processing capabilities to the 704 so people could experiment with checkers or music synthesis. The 704 was explicitly designed for military applications during the Cold War, sold as uniquely capable of the rapid and precise numerical calculations needed to simulate the trajectories of missiles and satellites.

In the early 1950s, Dantzig had moved from the Pentagon to the RAND Corporation, a think tank for military planning and psychology research, founded by the Army Air Forces and Douglas Aircraft, that had become an independent nonprofit in 1948. In chapter 3, I'll discuss RAND's obsession with war games. But they also were invested in computing, and Dantzig moved there to facilitate the development of better codes for linear programming. With collaborators at RAND, Dantzig significantly scaled up the solution of linear programs. On the 704, they were able to solve problems with a few hundred constraints and "virtually unlimited" variables.

IBM ensured that the 704 could handle the computations needed to optimally plan the trajectories of rockets and satellites, problems in aerospace that fell under the purview of what we now call *optimal control theory*. Optimal control theory marked a second phase in the development of modern mathematical optimization, and its development would recapitulate many of the lessons from linear programming.

Optimal control theory was a 1950s offshoot of the broader field of *automatic control*. Ominously named, automatic control studies how to build machines that automatically steer, regulate, and drive complex dynamical systems. We call these automated processes and the systems they drive *control systems*. Far from being a relic of the Cold War, control systems are invisibly embedded everywhere in our daily lives. In a modern car, cruise control is the obvious control system, taking over the pumping of the accelerator and brakes to maintain a constant speed on the highway. But dozens of invisible control systems are always running behind the scenes while you drive. For instance, the anti-lock brake control system prevents skidding, the air conditioning system keeps the cabin at a constant temperature, and the engine control unit manages the fuel injection system to meet emission requirements. Beyond the car, these hidden regulators undergird almost every aspect of our daily life.

Control systems are designed around the principle of feedback. The system to be controlled has an "input" and an "output." The control system is a dynamic system, with an output that adapts to its input over time. The output of the control system gets connected to the input of the system being steered, and the output of the system being steered is the input to the control system. As an example, think about keeping a room at a desired temperature by controlling a heater with a thermostat. If the temperature in the room is too low, the thermostat can output a signal to the input of the heater, turning the heater on. If the room gets too hot, the thermostat can change its output to tell the heater to turn off. These two systems, connected in feedback, then regulate the comfort of your room.

What would it mean to optimize a control system? Control design can be thought of as policy optimization. The control system needs a policy to compute its output based on the inputs measured from the steered system and other relevant aspects of its environment. Using our car examples, an engineer might aim to minimize the skid of an anti-lock braking system, have the AC minimize blower speed while maintaining a comfortable cabin temperature, or minimize the emissions output from an engine with a fuel injection system. Just as computers let us solve logistical policy optimization problems with linear programs, computers opened the door to *optimal* automatic control.

Also as in linear programming, the goals of optimal control problems are encoded as mathematical formulae. The constraints are the mathematical equations describing how the system and controller evolve in time. Each constraint describes what happens when the system is in a particular configuration and the control system responds with a particular action. A designer can also specify constraints on the inputs and outputs themselves, perhaps only allowing low-power inputs, or describing regions of space for a vehicle to avoid. Specifying a complex system with multiple inputs and outputs is straightforward in the modeling language of optimization. If we had a home with multiple thermostats, heating zones, and rooms to heat, this information could all be incorporated into the specification of the optimal control problem.

The first step along the path to optimal control was the invention of *dynamic programming* by Richard Bellman at the RAND Corporation in 1952. Though the name was clearly a riff on Dantzig's linear programming, dynamic programming was something entirely different, providing a general solution technique for the problem of optimal control. Dynamic programming is based on a deceptively obvious observation called "Bellman's optimality principle." The principle is easiest to describe in terms of planning routes of travel. Suppose you want to find the shortest drive from San Francisco to San Diego. Then if Los Angeles is on this shortest route, the segment of the journey from Los Angeles to San Diego is also the shortest route from Los Angeles to San Diego. That is, if the shortest route from A to B contains C, then the path from C to B in this route must be the shortest path from C to B. Using this idea, Bellman showed how you could solve optimal control problems by figuring out what would happen at the very end of the optimal plan and then working your way backward to the beginning.

Bellman's dynamic programming would become a centerpiece of computer algorithms, applicable in a variety of contexts beyond optimal control. Among its many applications, dynamic programming can find shortest routes, match patterns in sequences, and allocate resources. Because it was so generically applicable, the main drawback of dynamic programming was that it was often wholly impossible to compute optimal policies, even with a computer. Dynamic programming gave a path to solving optimal control problems, but it would only be a few simple instances that would prove influential.

In the late 1950s, Rudolf Kalman made dynamic programming practical for control by finding an important special case where the optimal policy was efficiently computable with computers of the day. Just as Harold Hotelling observed, the world engineers aim to control is complex and nonlinear. However, if an engineer is lucky enough to find a system where the physical laws can be modeled as linear equations, control design becomes simpler. One famous case where physics is linear is when Newton's laws of motion apply: Objects move proportional to how much force is applied to them. Once they move, they stay in motion, and all actions have equal and opposite reactions. Newton's laws aren't useful for modeling every mechanical control problem, but they are good approximations to objects freely traveling in the air. In this case, the control signal could be the force applied to an aircraft by a propeller or engine, and the goal might be to move the aircraft along a specified trajectory. For such linear systems, Kalman found a straightforward calculation that would compute the optimal policy using dynamic programming. This meant that if engineers could find reasonable regimes where their systems looked linear, they could optimally control their behavior. While computing these optimal policies by hand would be daunting, the sorts of calculations required mirrored those required to solve Dantzig's simplex method, and hence the computers of the day could rapidly find optimal controllers.

With the influence of Bellman and Kalman and complementary advances by Russian control theorists under Lev Pontryagin, the 1960s saw rapid advances in the application of optimal control. Complex automatic control tasks could be posed as policy optimization problems, and computers could solve for the optimal policies. For example, we could ask for a control policy to generate a particular maneuver of an aircraft. Or we could look for the most fuel-efficient way to move a vehicle from point to point. Kalman also showed how the problem of estimating the configuration of a dynamical system was complementary to the problem of controlling it. Using dynamic programming, he invented the *Kalman filter*, which was deployed on every Apollo mission to the Moon, tracking the lunar spacecraft and coordinating its maneuvers. Optimal control created a new modeling language for automatic control design, and it has played a pivotal role in the discipline since its inception.

One of Kalman's lesser-known papers highlights how the growing power of computers shaped his work.[18] Kalman worked as an engineer at IBM from 1957 to 1958. During that time, he helped bring optimal control to chemical engineering. In research with IBM colleague R. W. Koepcke, Kalman laid out a program to apply computers to optimize chemical reactors.

Flows of various chemicals fed the reactors, and the control system could adjust the flow rates of each chemical species. The control objective was to create desired concentrations of chemicals in the reactor as quickly as possible.[19] The constraints were the nonlinear dynamics of chemical reactions. Additionally, a constraint on the chemical flows was that the amount of material in the reactor had to be roughly constant over time. And the chemical flows themselves had to be constrained to be neither too fast nor too slow.

Despite the complexity of the problem, Kalman and Koepcke managed to find a reasonable approximate solution that worked in a computational simulation. Though the system was nonlinear, they designed a control policy that assumed it was linear. With this assumption, they could find the optimal control policy using dynamic programming. This dynamic programming solution could be simulated on any machine that supported floating point calculations. In their simulated reactor, this crude approximation was sufficient to quickly yield the desired chemical outputs. They wrote, "The role of designing a digital computer in real time control consists essentially of digesting large amounts of information obtained from the primary measuring instruments and then calculating as rapidly as possible the control action to be taken on the basis of these measurements." By assuming a linear policy, they could compute such policies in a timely fashion using the computing available at IBM.

At the end of their report, Kalman and Koepcke clearly and carefully list the many limitations of this approach of treating

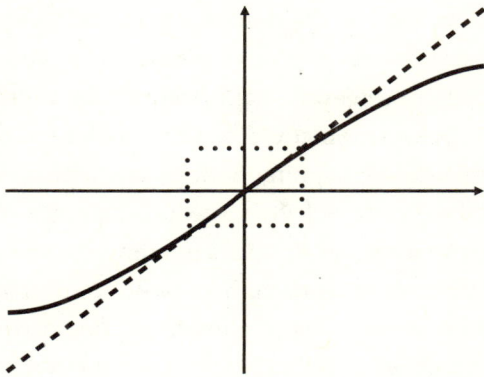

FIGURE 3. Approximating a nonlinear curve with a line. Inside the dotted square, the nonlinear and linear curves are nearly indistinguishable.

nonlinear systems as if they were linear systems. Importantly, they note that the only dynamic programming problems with practical solutions were those with linear dynamics. They suggest that more complex algorithms could be developed to tackle nonlinear dynamical systems, but "very little is known about this problem at present." However, little progress was made in the intervening decades. Efficient methods of dynamic programming for complex nonlinear systems remain a major open research challenge over sixty years later.

Since nonlinear problems seemed hopeless, Kalman and Koepcke also explored the limits of their linearity assumption. They noted that the linear approximation wasn't terrible around chemical equilibrium, and their simulations bore this out. Figure 3 shows a hypothetical curve relating variables in a nonlinear system. The dashed line represents a linear approximation of the system. In the enclosed box, the dashed line closely aligns with the nonlinear system. Kalman and Koepcke found that as long as the linear approximation of a system had sufficient overlap with the nonlinear system it was approximating, the linear approximation could be used in place of the nonlinear equations when designing policies. Kalman and Koepcke argued that if the system drifted into a configuration where the linear approximation was invalid, a new approximation around the new configuration could suffice for policy design.

Kalman and Koepcke's approximate solution of optimal control problems, called *linearization*, became standard practice in control design. Their idea of adapting to drift would later give rise to the practice of *iterative linearization*, where both linear approximations and control policies were computed in real time. Such iterative procedures remain state of the art for many robotic control systems today.

At this point in the chapter, Kalman and Koepcke's reasoning should now be familiar. They attempted to solve a complex control problem with a computer and quickly encountered computational roadblocks. By making various approximations of reality, they found a system they *could* solve with a computer. And then they proposed a variety of contexts and applications where engineers could force their approximation to be nearly true. The design principle of forcing reality to match optimal engineering calculations had become standard practice.

For twenty years after the invention of the simplex method, rapid progress in optimization algorithms continued, coupled with rapid progress in computing. Most of the major algorithms in logistics today, whether they be for optimizing supply chains or training neural networks, were developed between 1947 and

1969. On the hardware side, the transistor was invented in 1957 and quickly replaced the vacuum tube as the core computing building block. Shortly thereafter, in 1961 IBM released its first transistor-based supercomputer, the 7030. In line with their memorable naming conventions, IBM added an extra digit to the model numbers to indicate transistor-based computers. The 7030 was considerably more powerful than the 704. Whereas the 704 could perform 12,000 floating point operations per second, the 7030 could perform over 700,000.

Faster computers with more memory enabled more complex software. To tackle this newly available complexity, programming *languages* were developed. With a programming language, an engineer could declare what the computer should do at a level more abstract than individual arithmetic operations. Indeed, it is with the introduction of programming languages that the term "computer programming" takes on the connotation with which we are most familiar today. Programming languages enabled programmers to encode large transfers of bits from different parts of memory and declare entire steps of algorithms like the simplex method in a single line rather than a pile of punch cards. The earliest such language, FORTRAN, was developed at IBM in 1954.[20]

With the ability to write machine code at a more intuitive level, the sophistication and variation of mathematical programming rapidly mushroomed. Dantzig himself pioneered a new form of mathematical programming, *stochastic programming*, to design optimization rules that could adapt to uncertain future demands.[21] Researchers chipped away at the linearity assumptions of linear programming, finding ways to solve problems posed with nonlinear objectives[22] or when the solutions had to be discrete integer values rather than fractions.[23] These discrete-valued problems gave rise to modern computer algorithms like network flows, assignment problems, and shortest paths.

These new problems were specific examples of *nonlinear* programming. "Linear" is an easily defined term. Just as was the case when Harold Hotelling used the word to chastise Dantzig, the term nonlinear here is simply defined as the negation of linear. Any optimization model that is not linear is technically a nonlinear program. While it might be easy to model problems using nonlinear programming, solving them was another matter. I listed a few specific examples with solutions in the previous paragraph, but the scope of problems that could be modeled using nonlinear programming was impossibly large.

This overgenerality of nonlinear programming had practical consequences. Dantzig's simplex method is guaranteed to find *optimal* solutions for linear programs. Not only are the solutions of maximal value, but the algorithm verifies that the solutions are optimal. In stark contrast, no one could find a way of testing if a policy was optimal for general nonlinear programs. You could model a much wider variety of systems and costs with nonlinear models, but the price was that the beautiful certainty of linear programming had to be abandoned.

In response, nonlinear programming settled for finding *improved* policies rather than optimal ones. Even if you couldn't guarantee a particular solution was best, often you'd be satisfied with a *better* solution than the one you started with. This is the idea behind *local optimization*. If you have a candidate policy, are there algorithms to find policies that score better on your objective than the candidate? Perhaps you can try a policy similar to your candidate and see if it has a higher value of your objective function. And then you could continually iterate and refine the solution until you are unable to make further improvements.

The key ingredient for a local search algorithm is a candidate policy to start a search. From this candidate policy, the next step would be to find a nearby policy with a slightly higher value of the objective function. Such nearby policies could be found using an idea that dates back to the start of calculus. If a policy is improvable, there is always a perturbation, called the *gradient*, along which the policy can be improved. And the gradient can be calculated using rules from first-year calculus. Dozens of algorithms were invented in the 1960s, which proceeded by computing gradients and then moving along them. Famous examples included Rosen's gradient projection algorithm,[24] Fletcher and Reeves's nonlinear conjugate gradient method,[25] and Polyak's Heavy Ball Method.[26] People even took this a step further, approximating functions with quadratic rather than linear functions. The most famous of these methods, BFGS, is an initialism for the names of four researchers who simultaneously discovered the algorithm: Broyden, Fletcher, Goldfarb, and Shanno.[27] In the opposite direction and using even cruder information about the function, methods such as the Nelder-Mead method[28] or Fletcher's variable metric method[29] approximated gradients allowing for even more general functions to be optimized. All of these methods remain staples of today's optimization software packages.

Many of these local search methods found immediate applications in aerospace (be they for missiles, airplanes, or satellites), as the competition with the Soviet Union remained fierce. For instance, to design trajectories for rockets, Arthur Bryson combined classical mechanical techniques to analyze motions of objects with computational techniques to compute the resulting trajectories under various parameter settings.[30] He eventually found an elegant way to compute very complex trajectories that used dynamic programming to numerically compute the effect of small variable changes to the resulting trajectory. He called this a "steepest ascent method in the calculus of variations."[31] Today, this algorithm is called *backpropagation* and is the main algorithm used in training neural network models for artificial intelligence applications.

Since the very start of computing, there has been an insatiable demand for faster computers. No matter how large a problem a computer could solve, computer users would always demand the ability to solve something bigger. Computers needed faster processors, larger memories, and larger storage so we could solve larger problems. This dogma was good for computer manufacturers, as it kept demand for their new machines high. And it was great for computer scientists, who could always add a line item budget for a bigger computer to their grant report.

Computing speed initially grew at impressive rates. The IBM 7030 was around 60 times faster than the IBM 704. But IBM had actually overpromised, and consumers expected the 7030 to be over 100 times faster. In time for the Apollo Program, IBM had switched their numbering scheme and released the System 360–91, specially designed for the Houston Mission Control Center. This new supercomputer was over 20 times more powerful than the 7030. From 1954 to 1969, computer capabilities were doubling around every fourteen months.

With the Moon landing, however, computing speed began to plateau. One reason was that the unified pressure of the space mission had focused the military-industrial complex with a common task and corresponding need for faster computers. The United States having more or less won a major phase of the

Cold War by putting men on the Moon, the subsequent decade would inevitably be less focused.

But beyond the geopolitics, another reason for the stall was that early computers were complex and temperamental physical systems. IBM and other computer manufacturers would wire together these giant mainframes with a heterogenous assortment of parts that sprawled across giant server rooms. Each wire was a potential source of failure. Computers of the late 1950s could only run for about twelve hours before they would fail and require maintenance, and such maintenance would often require dozens of technicians and take a whole workday.

The rapid doubling of the power of mainframe technology cooled off, and every remaining incremental gain came with new engineering hurdles. Supercomputer speed slowly plateaued in the seventies and eighties. How did we get over this plateau? The answer was that the computer itself needed to be treated as a system to be explicitly optimized. And it would be optimized not only to be faster but also to be more reliable. Turning the developed tools of mathematical optimization back on the computer led to the greatest technological revolution humankind had ever wrought.

The mantra that scaling must occur for the computing industry to proceed is usually attributed to Gordon Moore, one of the founders of Intel. While employed at Fairchild Semiconductor in 1965, Moore wrote an opinion piece for the magazine *Electronics* entitled "Cramming More Components onto Integrated Circuits."[32] This article is most famous for predicting that the density of transistors in *integrated circuits* would double every two years. On computer mainframes in the 1950s, vacuum tubes were wired together in massive modules with many exposed parts and many potential places for manufacturing failures. By the early 1960s, technologies emerged that could build such modules as single devices. Using novel manufacturing techniques, multiple components could be hardwired together as part of the assembly process to form an integrated circuit.

Integrated circuits were a revolutionary change. Rather than a large circuit stitched together in open air, integrated circuits could be laid out in a tiny space and encased in unbreakable epoxy. Machine operators would never need to worry about thousands of fragile tiny wires and other exposed technology. Once a chip came off a production line, technicians could run a battery of tests on it and, if it passed, be confident it would function for years.

In hindsight, predictions about the technological future often seem laughably optimistic. But every prediction made in the introduction of Gordon Moore's 1965 article came true. Notably, he proclaimed, "Integrated circuits will lead to such wonders as home computers—or at least terminals connected to a central computer—automatic controls for automobiles, and personal portable communications equipment." The integrated circuit was the key to moving computation out of walled-off rooms in specialized labs and into every corner of physical reality.

This prediction has been colloquially used to assert that every desired feature of computers doubles about two years or so. But Moore was explicitly focused on the integrated circuit. And it's remarkable that given the situation in 1965, Moore's prediction was conservative. The early days of computing saw "doublings" every 12–15 months. Moore's prediction of doubling every two years may have cast him as relatively bearish on progress.

The slower doubling rate meant the impact of integrated circuits was initially modest. For many years Intel mostly made memory. But the compounding of exponential growth always wins out. With the 1985 release of Intel's 386, the home computers promised by Moore were over twice as fast as the best supercomputers from the 1960s.

In his prophetic *Electronics* article, Moore proposed, "Perhaps newly devised design automation procedures could translate from logic diagram to technological realization without any special engineering." This dream was realized in the 1970s by applying concepts of mathematical optimization to the design of integrated circuits.

Integrated circuits were planned out like cities. Transistors were houses located at various addresses and connected by complex routes of printed wires. Planning out how to place and route transistors to meet the desired logical goals of a computer was a daunting optimization problem.

The local optimization methods that I have described thus far used calculus to find directions of improvement. The methods of the 1960s required functions to vary smoothly as parameters were changed. But what if things didn't smoothly vary? What if small changes in a configuration would produce huge changes in the value? Such *non-smooth* problems were a hot research topic in the 1970s.

Chip design offered some challenging non-smooth optimization problems. Small movements of transistors would require large changes in route plans. Placing the transistors such that the resulting routes were shortest was easy to pose as a problem of non-smooth optimization, but it was a problem without a good solution.

The early designs of the 1970s embraced overconstraining possibilities. One of the first patterns to emerge was *standard cells* for design. Integrated circuits have common patterns used in many different designs. For example, processors would use logic gates that could compute functions like "Are either of the inputs 0 or 1?" Standard cells were fixed designs for such gates. Like in programming languages, these abstractions simplified design and allowed designers to think at a higher level about what they wanted on their chips. Like how LEGO blocks can build spaceships and race cars, intricate circuit complexity could be synthesized from these standard cells.

Standard cells simplified the design of basic circuits, but assembling standard cells into chips was still not automatable with optimization technology. That would change in the early 1980s with the work of computer scientist Alberto Sangiovanni-Vincentelli. Sangiovanni-Vincentelli was trained in mathematical optimization and interested in local search techniques. He moved to Berkeley in 1975 and applied his non-smooth search techniques to improve SPICE, an open source circuit simulator developed at UC Berkeley and widely used in the semiconductor industry. After a sabbatical at IBM, Sangiovanni-Vincentelli became interested in a new search technique inspired by physics, called *simulated annealing*. Annealing is slowly cooling a hot material to end in a strong state. For example, bending hot metal and cooling it drastically can result in a brittle final product, but cooling it slowly while bending it leads to a better structure. Simulated annealing uses this as a metaphor to describe how optimization should work: The algorithm maintains a "temperature" as it searches. When the temperature is "hot," simulated annealing searches wildly for potential good policies.

The algorithm reduces the temperature as optimization progresses, and policies are only slightly perturbed in search of extra improvement.

Simulated annealing probabilistically decides between staying with a current policy or perturbing the policy. Given a set of rules governing the movement from one policy to another, it will try a move, then evaluate the configuration. If the new policy is worse than the old one, it will sometimes accept the move in hopes of moving to a faraway configuration with a better objective function value. In simulated annealing, the virtual "temperature" parameter governs whether or not to accept these worse policies. By starting at a high temperature, more diverse policies are considered. As the temperature is lowered, simulated annealing focuses on final refinements of the current policy.

Simulated annealing was hard to understand and analyze in generality. Sangiovanni-Vincentelli found cases in routing and layout of chips where he could understand the behavior of simulated annealing and provably find reasonable solutions. Sangiovanni-Vincentelli and his students applied these optimization insights to circuit layout, developing a new software package called TimberWolf for laying out standard cells, YACR (Yet Another Channel Router) for routing wires, and MOSAICO for laying out higher-level cells. At the same time, Sangiovanni-Vincentelli had become interested in Moore's vision of converting high-level descriptions of circuits into layouts. With his students, he designed methods to go from a high-level description of circuits called *register transfer level* (RTL) to a sequence of logic gates. RTL serves a similar role to the programming languages I described earlier, letting designers focus on more abstract levels of circuit design and think about how digital bits would flow through a circuit.

All of this work built up a software suite that could design chips with significant complexity. Sangiovanni-Vincentelli and his Berkeley collaborators had been working closely with industry and decided to form a consulting company to further the development of their software. They founded SDA in 1983.

In 1984, Intel decided to adopt SDA's tools for the design of a new processor. That processor was released in 1985 with the model number 80386. The 80386, usually just called the 386, became the de facto processor of the personal computer in the 1980s. The 386 opened the door for companies like Compaq to offer personal computers to consumers without manufacturing the chips. Other companies followed suit, and dozens of manufacturers soon offered "PC clones" with 386 microprocessors. The 386 cemented the foundation of Intel as the preeminent American manufacturer of personal computing microprocessors.

The 386 consisted of 275,000 transistors, nearly ten times the number of Intel's first successful microprocessor, the 8086. The 8086 had been designed by hand by a team of sixteen designers who worked for two years on the final layout. The 386 was designed with computers, using the code of SDA. SDA would later merge with a company called ECAD, and the merged entity was dubbed Cadence. Cadence remains the preeminent developer of industry tools for chip design.

Reflecting on the impact of computer-assisted design on the computing industry, Sangiovanni-Vincentelli said:

> IC design was largely carried out with few expensive tools running on large mainframes or specialized graphic processors. The design process was slow, often manual and error prone for the lack of verification and analysis tools. The electronic industry was confronted with serious productivity

problems. We changed the situation: the introduction of layout editing and simulation on inexpensive general-purpose workstations as well as the introduction of algorithms for the automatic layout of gate-array and standard-cell chips yielded at least one order of magnitude productivity improvements. It made it possible to develop semiconductor chips that were working correctly the first time as opposed to the several manufacturing runs needed earlier.[33]

Constraining the process and design and then automatically optimizing led to unprecedented progress in computing. The hyper-controlled world of the integrated circuit lets designers build more and more sophisticated devices on single chips, combining the chips designed in the past into greater and greater complexity. If the 8086 was a small city, the 80386 was a megalopolis. And today's multicore monstrosities from NVIDIA that mine Bitcoin or compute AI models are endless sprawling combinations of multiple megacities.

―――

Over the course of seventy years, the codependent development of optimization algorithms and computers brought us from warehouse-sized mainframes that could perform 10,000 computations per second to tiny devices that fit in our pocket and execute 2 trillion operations per second on the graphics processor alone. People across the world have access to handheld computers that are millions of times more powerful than those available when Moore made his visionary predictions.

What we do with these trillions of computations might not be what the optimizers of the fifties envisioned. It is true that smartphones let us do wonderful things, like have video conversations with family on the other side of the planet. But I would be remiss not to dwell for a moment on how most supercomputing is devoted to mindlessly scrolling through video content on social media platforms. We simultaneously observe the most impressive engineering in history and realize that most of this technology makes money by surveilling our every action to serve us targeted advertising.

This leads us to a crucial lesson about what can and cannot be optimized. Moore observed in 1965 that the laws of physics suggested we could make integrated circuits smaller, faster, and more ubiquitous. Accepting the challenge, chip designers developed software design tools to enforce a rigid set of rules upon the organization of an integrated circuit. Playing by these rules enabled chip designers to build microscopic digital cities that process unfathomable amounts of data. Thanks to decades of industrial will, today's computer chips are highly optimized, overdesigned physical systems of breathtaking complexity. Computer chips were easy to optimize precisely because they are small, closed worlds with simple rules and constraints that wall them off from the unpredictability of the world around them.

Outside the computer chip things were far less tidy. The limits of optimization became strikingly apparent in the physical by the end of the Cold War. After fifty years of success in military applications, control engineering was developing breakthroughs in space exploration, aviation, and energy. However, it also contributed to a series of disasters across these sectors, which Honeywell aerospace engineer Gunter Stein argued should give the community pause. At the 1989 IEEE Conference on Decision and Control, Stein delivered a gripping plenary on how our idealized models can lead us astray when we try to use them to optimize complex physical systems.[34] He argued that understanding the dangers of modeling uncertainty was critical to pushing the field forward.

Stein noted that though we can design control systems that work for complex nonlinear systems, these systems are only *locally* controllable. As we've discussed in this chapter, control systems are designed to force systems to behave as if they were linear so they can be properly controlled by computational tools. However, Stein pointed out through several examples how small, unplanned disruptions can quickly drive a system into an uncontrollable state, at which point disaster could occur.

Stein's closing example riveted the crowd. He described how the Chernobyl disaster of 1986 resulted from not respecting the limits of engineers' idealized models.

> Whether we choose to recognize it or not, control played a major role in that accident. The plant's hardware did not fail. No valve hung up, no electronic box went dead, and no metallurgical flaw caused a critical part to break. Instead, the reactor control system systematically drove the plant into an operating condition from which there was no safe way to recover. This is true, at least, if we count the control system's hardware, its human operators, and its operating policies as part of the system.

Stein concludes his lecture with a call to action for control engineers.

> This reactor control application, as well as the airplane applications I talked about earlier, illustrate that society does indeed permit control engineers to operate dangerous systems. The number of such applications increases steadily. Not all of them have such dramatic consequences as Chernobyl, but they are dangerous nevertheless.

In 1989, these systems seemed destined to connect us to the rest of the universe. Stein warned that if we wanted to push our engineering further, it would require great care and investment, but also unavoidable danger. Some might interpret Stein to be warning us that "it's not worth it." Control theory has its limits, and though it had delivered more than we could have imagined, perhaps it had delivered all that it could.

What would the world look like if we decided not to pursue dangerous control applications but instead chose to work with the ones we had and make them as safe as possible? The answer is that the physical world would probably look a lot like the one we live in today. Control systems are undeniably everywhere—in our heating systems, computers, cars, and airplanes—creating a hidden infrastructure that connects our world. But we have the same number of nuclear plants running today as in 1989. We are not sending more people into space. We aren't building supersonic commercial jets; commercial aviation is, on average, slower today than in 1989.

On the other hand, aviation is much safer today. Deaths on American commercial aircraft are exceptionally rare. Flight prices are low, and air travel is more accessible than ever (though you'll have to pay extra to avoid having to take off your shoes and belt). Control and feedback still underlie all aspects of the infrastructure of our hyperconnected world, but its aims are decidedly more modest than those imagined by the futurists of the 1950s and 1960s.

By the 1990s, the Cold War had ended, and intrepid engineers realized they could make a lot more money from the micro world than the macro world. The internet age roared in.

It was safer and more profitable to convince people to click on ads than to design control systems for nuclear power plants or next-generation air travel.

In addition to being safer than nuclear power or space travel, microprocessor scaling points to some of the fundamental prerequisites for optimization. Optimization takes capabilities we have and makes them better by overconstraining them. This works well for improving the infrastructure that we already have. The initial condition from which we optimize is critical. We can't use engineering optimization to will the nonexistent into being. We can only take what we have before us and make it better for some narrow and specific notion of better. For this reason, Gordon Moore's predictions from 1965 look prophetic, but contemporaneous predictions about flying cars sound as outlandish today as they did then.

Let's compare and contrast some of the other technological predictions from the early computing era, to get a better grasp of which have been realized today and why. Recall that Moore wrote, "Integrated circuits will lead to such wonders as home computers—or at least terminals connected to a central computer—automatic controls for automobiles, and personal portable communications equipment." Already at the time of his writing in 1965, we had modern computer architectures, automatic vehicle control, and a robust knowledge of wireless communications. Moore just predicted that we could make these smaller, faster, and more ubiquitous.

In 1993, AT&T ran a series of TV commercials, directed by David Fincher and narrated by Tom Selleck, about the amazing technologies they would soon bring us. Showing a computer terminal at the library, Selleck asks, "Have you ever borrowed a book from thousands of miles away?" The ad shows a GPS system in the dashboard of a car and asks, "Crossed the country without stopping for directions?" "Have you ever sent someone a fax from the beach?" At the end, triumphantly, Selleck cheers, "You will, and the company that'll bring it to you is AT&T." The ad series highlighted wireless highway toll systems, automatic cashless kiosks for ticketed events, streaming movie services. Everything promised ended up being delivered (though not necessarily by AT&T). But again, this was because these technologies existed, and computers only had to get a little bit faster and a little bit smaller to make them a reality.

By contrast, in 1958, *The New York Times* published an article claiming, "The Navy revealed the embryo of an electronic computer today that it expects will be able to walk, talk, see, write, reproduce itself, and be conscious of its existence." This hasn't happened yet. And, never pausing to reflect, *The New York Times* continues to publish similar articles today.

In 1961, CBS produced a documentary called *Tomorrow* to commemorate MIT's 100th anniversary. The end of the episode "The Thinking Machine" features a monologue by Claude Shannon, whose work on digital logic, information theory, computational game play, artificial intelligence, and natural language processing had established him as a prestigious thought leader at MIT.[35] In the documentary, he proclaims, "I confidently expect that within ten or fifteen years we will find emerging from the laboratories something not too far from the robot of science fiction fame." Why did this not come to pass?

Part of Shannon's monologue contained an important clue to understanding what technologies would flourish. Shannon pointed out that even in 1961, "We have machines that will translate to some extent from one language to another. Machines that will prove mathematical theorems. Machines that will play chess or checkers sometimes even better than the men who designed them." All of these technologies have been vastly improved since 1961, but the seeds were there already. We just had to optimize the infrastructure on which they ran.

But to go a step even further, why were these sorts of "intelligent" technologies discovered in the 1950s? It is precisely because these particular applications that Shannon called out *could be reduced to optimization problems.* This remains true today. As I'll discuss in much greater detail in chapter 5, all our modern Artificial Intelligence systems are based upon optimization,[36] with large data centers devoted to running optimization algorithms to minimize the number of errors of statistical prediction models. In fact, these prediction-error-minimizing machines are based upon the very Navy computer embryo overhyped by *The New York Times*, Rosenblatt's perceptron algorithm. And they are trained with extensions of Bryson's algorithm from 1962. And yes, even today's captivating chatbots, heralded upon their arrival as an artificially intelligent revolution, are built by predicting the next word in sentences.

Shannon hoped for "a more general computing system capable of learning by experience and forming inductive and deductive thoughts." Such systems remain far out of reach, likely because life, experience, and thoughts are not the solutions to an optimization problem.

The computer age has taken World War II technology and made it smaller, more embedded, more accessible, and much, much faster—in one word, optimized. But the rest of the world can't be boxed off and optimized, for better or for worse. Human cities and societies are not microchips. They are messy and resist authority.

3

This Is Not Nam. There Are Rules

IN 1944, physicist John von Neumann and economist Oskar Morgenstern published an audacious manifesto, *Theory of Games and Economic Behavior*.[1] Von Neumann and Morgenstern proposed that we could formally capture the problems of economics in the language of mathematics. They aimed to provide "an exact description of the endeavor of the individual to obtain a maximum of utility, or, in the case of the entrepreneur, a maximum of profit." Proper formalization of economic problems could ignite a revolution similar to twentieth-century physics, completely changing our capabilities for understanding economic phenomena and enabling quantitatively impactful policy. And perhaps through this new lens, we could establish a formal understanding of all human behavior.

Von Neumann had done much of the mathematical groundwork for this book in the 1920s in an article called "Zur Theorie der Gesellschaftsspiele."[2] This work, whose title translates to "On the Theory of Parlor Games," lay dormant for decades. It wasn't until the Second World War that an interest in game theory reemerged. Von Neumann was broadly active in the war project, and helped advise military strategists using some of his concepts from his seminal paper. In discussions with Morgenstern, he realized that much of this work could be applied broadly to economics. But the goal of *The Theory of Games* was more ambitious than what von Neumann had set out to do twenty years earlier.

Given the incredible advances in physics in the preceding two hundred years, it was impossible for von Neumann to approach the problems of economics without a little bit of ego.[3]

> It is not that there exists any fundamental reason why mathematics should not be used in economics. The arguments often heard that because of the human element, of the psychological factors etc., or because there is allegedly no measurement of important factors, mathematics will find no application, can all be dismissed as utterly mistaken.

Despite its hubris, von Neumann and Morgenstern's project was remarkable. Through a good deal of sophisticated mathematics, they formalized the mathematical notions of the economic concepts of *utility* and *strategy*. Utility was a function that measured a person's satisfaction with the potential outcomes of economic engagements. A strategy was a rulebook used by an individual to decide how to act in any given encounter. Von Neumann and Morgenstern proposed that individuals seek to maximize utility by finding optimal strategies. With this model of people, they asserted that "[t]he typical problems of economic behavior become strictly identical with the mathematical notions of suitable games of strategy." Von Neumann and Morgenstern found that all human interactions could be modeled as playing games under appropriate abstraction.

Von Neumann and Morgenstern worked to model human interaction through different types of games, enumerating possibilities by the number of players. One-player games are a special type of policy optimization problem considered in the last chapter. A utility function characterizes the value of each outcome at the end of the game, exactly like the objective function in mathematical optimization. Each strategy consists of a set of choices taken by the policy over time, and the constraints of the game specify how each choice leads to the next stage of the game. The optimal policy would be the strategy that resulted in the highest possible utility. But these games that reduced to central planning weren't interesting to von Neumann and Morgenstern. They didn't seem to yield any economic insights outside of "Robinson Crusoe" economics, where a single agent, cut off from the world, had to rely on only the resources on a desert island. Though von Neumann and Morgenstern conceded that these games could potentially help regimes of authoritarian central planning, they only devoted two pages of their book to one-player games.

Two-player games are where the theory becomes interesting. Von Neumann and Morgenstern formalized the idea of a *zero-sum game*. In such a game, whatever utility is received by the winning player is lost by the loser. For example, they describe the game "Rock, Paper, Scissors." Think about the simple winner-takes-all version of this game without any repetition. Each player has to choose rock, paper, or scissors. Rock will beat scissors because rocks can smash scissors. Scissors, on the other hand, cuts right through paper. And paper apparently can cover rocks, so that's a win. If both players choose the same move, we declare a draw. In any round of this game, either one player wins and gets a point, or both players tie. We can make this a zero-sum game by awarding 1 point to the winner, taking a point away from the loser, and awarding zero points if there is a draw. We can sum up the outcomes in table 1.

TABLE 1. Outcome table for Rock, Paper, Scissors

	P2 plays rock	P2 plays paper	P2 plays scissors
P1 plays rock	0 for P1, 0 for P2	−1 for P1, 1 for P2	1 for P1, −1 for P2
P1 plays paper	1 for P1, −1 for P2	0 for P1, 0 for P2	−1 for P1, 1 for P2
P1 plays scissors	−1 for P1, 1 for P2	1 for P1, −1 for P2	0 for P1, 0 for P2

Von Neumann and Morgenstern also cast chess in their game theoretic framework. At the end of a game of chess, the only outcomes are a win for white, a win for black, or a draw. But chess is distinguished by being a game of *perfect information*. Each player observes every move in the game, and nothing about the board is hidden from each other. When making the next move, a player with the perfect strategy book can look up the history of everything that has occurred thus far and then make the corresponding move in the strategy book. Since the game has perfect information, there is no ambiguity in which move to take. In this sense, von Neumann and Morgenstern declared the game of chess to be trivial: If both players play with the optimal strategy, either white always wins, black always wins, or the game always ends in a draw. Von Neumann and Morgenstern don't tell us what the optimal strategy *is*, of course. Computing this optimal strategy would turn out to be more difficult than expected.

A game with *imperfect* information is poker, and von Neumann and Morgenstern devote nearly a chapter to it. In poker, each player holds cards that the other players can't see. Even in a game with two people, this makes an optimal strategy for poker challenging. If our opponent is continually raising their

TABLE 2. Outcome table for Matching Pennies

	P2 (odd) plays heads	P2 plays tails
P1 (even) plays heads	+1 for P1, −1 for P2	−1 for P1, +1 for P2
P1 plays tails	−1 for P1, +1 for P2	+1 for P1, −1 for P2

bets, does this mean they have an unbeatable hand or that they are bluffing? Because such bluffing was possible, it was no longer at all clear how to write the optimal playbook. If a player always bluffed, and their opponents knew it, the bluffer would lose their shirt on bad hands. However, if a player never bluffed, their opponents would quickly drop out whenever the player raised their bets. So some sort of strategy that selectively mixed in some bluffing was necessary. Von Neumann and Morgenstern showed that these *mixed strategies* could be achieved by *randomly* choosing one of the two options with appropriate probability.

Mixed strategies are easier to describe in the simple game of penny matching. Two players, player 1 ("Even") and player 2 ("Odd"), each place a penny on the table. If both are heads or both are tails, player 1 gets a point and player 2 loses a point. If one is heads and the other is tails, player 2 gets a point and player 1 loses a point. We can again summarize the possible outcomes in table 2.

If player 1 knows that player 2 is always going to play heads, player 1 will also always play heads to keep them "even." But if player 2 knows that player 1 is going to play heads, player 2 will play tails to make them "odd." In either case, player 1 and player 2 are best advised to have some hidden strategy. Compare this to chess: If both players were playing optimally, each would know the other's strategies and yet would still not change course. Sticking to the optimal playbook is the best option for both of them.

What's the solution for matching pennies? It turns out there *is* an optimal strategy to play so that even if you tell the other player your strategy, they won't change what they are doing. The best you can do here is to flip your coin and let randomness rule. If player 1 plays a coin flip, then the best player 2 can do is win half the time, as there is no way to guess the outcome. The same goes with the roles of the players reversed. So the optimal strategy is to play completely at random, and, in such a case, both players win about half of the time.

The random nature of matching pennies led Paul Samuelson to declare, "Chess is trivial, matching pennies is not." The optimal strategy for chess is to follow, deterministically, the optimal playbook. Even if black knows what white's optimal next move will be, white will still take that move because it is the singularly optimal move.

Calling games like chess "trivial" reveals much of the perspective of game theory. On the one hand, the fact that games like matching pennies required randomness was shocking in the 1940s. Von Neumann and Morgenstern had proven that to be rational (that is, mathematically rational), one necessarily had to be *unpredictable. Randomness* was rational. This revelation was utterly unexpected.[4]

On the other hand, everyone also knows that chess is an incredibly challenging game to master, and even grandmasters do not have access to the idealized optimal playbook. What von Neumann and Morgenstern's theory left out was *computation*. Just because an optimal strategy exists, does that mean we can find one? Even finding the optimal strategy for player 1 in games where player 2 only has one move is a computational challenge if player 1 has too many moves at their disposal. Indeed, any mathematical optimization problem could be reformulated as a one-player game. As we saw in the last chapter, mathematical optimization in its full generality is computationally difficult.

TABLE 3. Comparing the key properties of some two-player, zero-sum games

	Information	Strategy	Computable
Rock, Paper, Scissors	Imperfect	Mixed	Yes
Matching Pennies	Imperfect	Mixed	Yes
Chess	Perfect	Pure	No
Poker	Imperfect	Mixed	No

Table 3 sums up the differences between the two-player games covered by von Neumann and Morgenstern. Games like Rock, Paper, Scissors and matching pennies were imperfect information games with easily computable strategies. Chess was a "trivial" perfect information game, but no one knew how to write down its optimal strategy. Poker appeared considerably harder, an imperfect information game with an unknown optimal strategy.

If we can't find an optimal strategy because of computational limitations, can we find one that's good enough to beat everyone else most of the time? Mathematicians Richard Bellman and David Blackwell lamented that finding optimal strategies of games was a game in itself:

> The actual solution of any particular game is generally a matter of some difficulty, involving a combination of frontal attack, applying routine methods, and a type of mathematical ingenuity that has been described as "low cunning."[5]

These questions of computing optimal strategies would obsess researchers in the nascent field of Artificial Intelligence. From the perspective of policy computation, chess turned out to be anything but trivial.

What would an optimal strategy for chess look like? For each configuration of pieces on the board, a playbook declares which move to take next. Von Neumann and Morgenstern's work showed there had to be an optimal playbook. Every valid board had an optimal next move and hence a predetermined fate. Optimal play would result in a win for White, a win for Black, or a stalemate. But could you write that playbook down? Could you prove that white always wins if played optimally? Or that games should always end in a stalemate?

Claude Shannon would define a discipline in trying to answer these questions. He set out to try to find an optimal strategy for chess in the late 1940s. Shannon was interested in chess not because it had anything to do with economics, but because he loved the game and thought it would be a challenging puzzle to build a computer program that could play. In his words, "Although perhaps of no practical importance, the question is of theoretical interest."

Using some back-of-the-envelope calculations, he tried to determine how big the optimal strategy book for chess would be. Even his crude estimates led to unfathomable numbers. There are 64 positions on the chess board. In each position, there can either be a white piece, a black piece, or an empty space. Any chessboard will have at least 32 empty spaces. Each player has 8 pawns; 2 each of rooks, bishops, and knights; 1 king, and 1 queen. By just computing the total number of ways to arrange these 16 pieces, ignoring whether or not the board could be reached by legal moves, Shannon estimated there were about 10^{43} total board configurations. How can you even visualize such a large number? Humans currently produce absurd amounts of data, totaling around 10^{21} bits per year.[6] At such a rate, it would take more than 10^{22} years to write down the optimal code book for chess. That is a trillion times more than the age of the universe.

The optimal codebook was clearly not possible. Shannon then wondered if there would be a way to an optimal strategy using *game trees*. Game trees were developed by von Neumann and Morgenstern as a way to best describe the many forking paths down which players could go in a game. Every game tree has a *root*, the current position of the board. From the root, one can draw *branches* of every possible move accessible to the current player. This results in a collection of *nodes* of possible next boards. From these nodes, we can branch again, resulting in a collection of new nodes under each prior node. Continuing this process, the branching process eventually ends with *leaf nodes* where the game has reached a conclusion of a win for a player or a draw.

If you could build a game tree, finding the best move is "trivial" in theory. If both players played optimally, every leaf would have been reached by an optimal move. This means the previous move would have been optimal. So we can work our way back from the leaf nodes in the tree to the previous nodes, recording the score in the previous nodes. By climbing up the tree and counting the optimal moves, we eventually get back to our root node. But at this point, the branches are enumerated by the optimal next play, and the move has been computed.

The problem is that game trees are also impossibly large. For chess, Shannon estimated that there would be about 40 branches at any node. If he restricted his attention to games of at most 40 moves, this would mean that the full game tree would have 10^{120} nodes. I don't even have a good comparison for what 10^{120} means. There are only 10^{80} atoms in the universe. "The number of nodes in chess' game tree is equal to the number of atoms in one septillion universes." Does such a description make the number any more comprehensible? No. Once again, Shannon found that a brute force approach would require an impossible amount of computation.

Most of the games we play don't lend themselves to the small 2 × 2 tables of Matching Pennies or 3 × 3 tables of Rock, Paper, Scissors. Part of what is so compelling about gameplay is the dazzling complexity hidden behind games with apparently simple rules. From Shannon's calculations, it was evident that humans did not play chess optimally. Or, if they do, there would never be any way of *proving* anyone played optimally. Writing down the proof would require more bits than the universe had to offer. But Shannon was still intrigued by the problem: Just because we couldn't find the optimal solution, could we still write a program that could compete with a human?

To get around the complexity barrier, Shannon leaned on two tricks: First, it was clear that he couldn't enumerate all boards to a depth of 40. But enumerating all boards from the current board out to say, 3 moves in the future would only require computing about ten billion boards. Given the computing power at the time, this might have been one full day of computation. If he restricted attention to only modestly reasonable moves, there would be even fewer scenarios to evaluate.

But for most boards, three moves ahead would not result in a checkmate or draw. There had to be some other way of scoring the leaf nodes of this shorter game tree. Shannon decided one could use heuristic scores like "number of pieces" to rank the nodes in the smaller trees. With heuristic scores, the computer could compute a small game tree, look at all of the nodes, and pick the one with the best score, then trace its way back up the game tree to find the best move from its current position. Shannon called these scores "approximate evaluating functions." Today we call them *approximate value functions*. I will refer to them this way for the rest of the chapter.

These two techniques of Shannon's—bounded-depth tree search and approximate value functions—would form the foundations of computational gameplay.[7] For the next fifty years, they would be the basis of all successful computer chess programs. First, time would be spent developing clever approximate value functions using knowledge of chess. Second, chess programmers would look for innovative ways to search game trees as deeply as their computers would allow.

Unfortunately, the progress in chess was slow. In 1958, Alex Bernstein and his colleagues at IBM were able to make a chess-playing machine that could play with modest competence. Bernstein's program used Shannon's algorithm. It first restricted the number of moves to consider at any turn to 7. It used an approximate scoring function and looked 2 moves ahead to plan its move. This only required computing about 3,000 scenarios. Even with such restrictions and the most powerful computer available at the time, the IBM 704, it took eight minutes to determine its move.

In *Scientific American* in 1958, Bernstein lamented, "Even with much faster computers than any now in existence it will be impracticable to consider more than about six half-moves ahead, investigating eight possible moves at each stage." This would be the most modest of improvements over his current program. He then opined that perhaps "[a] more promising line of attack is to program the computer to learn from experience."[8]

In the late fall of 1948, Arthur Samuel faced the problem all academics face eventually: He was running out of grant money.[9] Samuel had joined the University of Illinois and led a project to build out a scientific computing effort at the University, in direct competition with efforts on the East Coast at MIT,

Harvard, the Institute for Advanced Study, and The University of Pennsylvania.

But the project coffers soon ran dry, and Samuel brainstormed how to fundraise. He had seen a talk by Shannon about his work on computer chess, but Samuel mistakenly thought that Shannon had already built a chess-playing computer. From this wrong information, Samuel inferred that "it ought to be dead easy to program a computer to play checkers." Checkers was an ideal candidate game. It was not only easier for people to play checkers than chess, but it seemed clear that it would be easier to program a computer to play checkers too. A checkers board has the same number of squares as chess, but far fewer different kinds of pieces with far fewer available moves. Using a similar count to the one Shannon made for chess, the number of different games of checkers was "only" 10^{20} as compared to the gargantuan 10^{120} for chess.

Samuel thought he might be able to build a prototype checkers player in less than a year and then enter it in the checkers world championship in Kankakee, Illinois, 75 miles up the road from the university. This would surely help him fundraise!

Samuel considered gameplay an ideal playground for investigating how to build computers that learned. In games, there was a clear and definite goal (i.e., winning the game).[10] The rules of games were definite, known, and never changed. Popular games were accessible by people, so the difficulty of the task could be appreciated and people would be impressed if the computer was a good player. It would be unambiguous if a computer program "worked."

The project didn't turn out as planned. He sketched out the prototype for a checkers algorithm and quickly realized he'd need a much larger computer than he envisioned and wouldn't have the budget to build it. Disillusioned with his progress and by the funding pressures of academia, Samuel left Illinois for a research scientist position at IBM the following year. There, he would work with the team designing the IBM 701. And he would design the 701 so it could assuredly master checkers.

At IBM, Samuel found another appealing aspect of games. Computer games were ideal programs to test whether computers were being designed with the appropriate capabilities. His logic designs for the 701 imagined what would be needed for his hypothetical checkers program. In his mind, games could be used "to test the proposed instruction set for its completeness and its effectiveness as a tool for expressing the operations that one would want the computer to perform."

As the 701 came into production, Samuel scheduled time to implement his checkers player. At IBM Samuel had worked on his checkers program for a decade, slowly experimenting at odd hours when there were spare cycles on a mainframe. Though the computer played competent matches as early as 1956, he only published a report on his work in 1959. IBM had apparently feared publicizing his work because it showed how humans were losing out to machines. Unlike today, this was considered bad PR in the 1950s.

Samuel's basic approach mimicked Shannon: he built checkers-specific heuristics for approximate value functions and then used tree search to find the best move. The main departure from Shannon was implementing what Bernstein had wanted for chess: Samuel programmed his player to *learn* the function that estimated the move values. Samuel's learning algorithm was visionary. He would provide the final piece of contemporary computational gameplay: self-play.

Samuel's learning program had two "players" he called Alpha and Beta. These two programs would play against each other.

Beta would use the current best strategy available. Alpha, on the other hand, would learn a new strategy over the course of the game. If Alpha won the game, Alpha's strategy was given to Beta. Alpha would continue to learn against the updated Beta in the subsequent games. If Beta won a game, Alpha was given a strike. After Alpha received three strikes, Samuel would reset parts of Alpha's program so that it would hopefully stop following such counterproductive directions. Samuel would hone his program for a decade at IBM, using late-night hours when technicians were at home to have Alpha and Beta play each other.

How did Alpha learn? It used an idea that we now call *temporal differencing* (though this term would not be coined until the 1980s).[11] At any point in the game, Alpha would be working with a mathematical model that assigned an approximate value to every possible move. Based on this model, it would take the move that seemed most advantageous. Beta would then respond. This would bring the board to a new configuration, and Alpha could use its model to estimate the value of the next move. The key to temporal differencing is that Alpha's estimation of the value of its current position would be different from its estimation one move ago. Because Beta had responded, Alpha received new information about its strategy. The difference of the estimated value of the current position minus its estimated value a turn before is called a "temporal difference." This temporal difference could be used to update the function that estimated the value of moves. Updating the value estimates every move, Alpha would end up with a very different function at the end of a game than from the beginning, and if the game was won, we might expect this new function to give a better strategy than the old one.

Samuel's program turned out to be remarkably competent. It defeated an excellent player, Robert Nealy, in 1962.[12] However, in 1966 when playing against master checkers players, Samuel's program lost all of its matches.[13] Still, it was certainly better at checkers than Bernstein's chess player was at chess. Samuel called his computer program a "machine learning" program, and he is credited for inventing the term.

The first machine learning system was built to play games. But could it be used to learn outside the rigid and clear rules of checkers?

At this point in the chapter, you might be wondering how, though we started with a discussion of economics, I've spent the majority of the time talking about board and parlor games. You would not be alone! Upon its release, economists and social scientists were enchanted by von Neumann and Morgenstern's theory of games. Herb Simon, a Nobel Prize laureate who would later strongly oppose the von Neumann and Morgenstern formulations, initially lavished the book with praise. He thought that the ambitious paradigm they set out for seemed to give a certain orderly, rule-based approach to postwar prosperity. However, Simon also remarked that much work was to be done. While the analogy with gameplay was alluring, Simon noted that no actual applications in sociology or political science were worked out in detail in the text. Though they made loose analogies to economic scenarios, all of the concrete examples of zero-sum games in the book were parlor games.

Von Neumann believed this absence was an artifact of experience. Though he had consulted with military leaders during World War II, he had not been able to propose actual game theoretic strategies. Shortly after *Theory of Games and Economic Behavior* was published, von Neumann set out on a path to move his theory to practice. In collaboration with the U.S. government, he gathered together talented mathematicians to apply game theory to military strategy.

The result was the founding of the RAND Corporation in 1948, which we first encountered as a patron of optimization research in chapter 2. RAND, an aptly named think tank focused on military research and development, was launched with funding support from the research offices of the Air Force and the Douglas Aircraft Company. Located in a nondescript office building on Main Street in Santa Monica, RAND would be the center of applied game theory research for the next two decades, and most of the famous game theorists either worked there full-time or visited regularly. The game theory research group was set up by von Neumann and employed future Nobel laureates Lloyd Shapley, National Medal of Science winner David Blackwell, and dynamic programming inventor Richard Bellman. The illustrious game theory group at Princeton, including Nobel laureate John Nash and mathematicians Harold Kuhn and Albert Tucker, were frequent visitors to RAND. Future Economics Nobel Prize winners Kenneth Arrow and Robert Aumann also spent time at RAND.

It seemed reasonable that military strategy might be formalizable as a zero-sum game between warring factions. Many detailed studies were done at RAND to counterfactually argue that particular campaigns might have gone better had optimal strategies been deployed. However, these studies were biased by the taste of the RAND research staff, most of whom were mathematicians.[14] Despite their best intentions, mathematicians tend to abstract concepts to places with little contact with reality. The game theory research at RAND was motivated by military applications, but in actuality most of the connections to battle were through metaphor.

Richard Bellman's game theoretic work with David Blackwell provides a great example of the strained metaphor between games and military strategies at RAND. Bellman and Blackwell laid the foundations of what would be known as "games of timing." Their first paper on the topic was called "A Bomber-Fighter Duel."[15] Their report begins describing a warfighting scenario "of a fighter, capable of firing a single rocket burst, attacking a bomber, which defends itself by firing intermittently." But by the next paragraph, the paper has devolved into integrals and other esoteric mathematics.

Blackwell and Max Schiffman later computed the optimal strategy of this hypothetical duel. I quote it here directly to give the reader a feel for the sorts of prescriptions that game theory provided.

> It turns out that, for fixed accuracies and values of the bomber and fighter, the nature of the strategies depends on the amount δ of ammunition at the disposal of the bomber. There are two critical amounts δ_0, δ_1, δ_0 greater than or equal to δ_1. If δ is less than δ_1, the bomber starts firing at a specified range at less than the maximum intensity and continues firing, with decreasing intensity, until the end of the engagement. If δ_1 is less than δ and δ is less than δ_0, the bomber starts firing at a specified range, greater than in the preceding case, fires at full intensity for a specified time, and with decreasing intensity for the rest of the engagement. In both cases, the fighter uses a mixed strategy, spread over the entire time the bomber is firing at less than maximum intensity, with a positive probability of firing at range 0. This probability decreases as δ increases. For δ greater than δ_0, the bomber fires at full intensity from the time the fighter comes within range until a specified time, and decreasing intensity

thereafter. There is a certain time, during the period in which the bomber is firing at full intensity, at which the fighter should always fire. Ammunition in excess of 80 is useful to the bomber only in case the fighter makes the mistake of waiting too long before firing.

These strategies required precise quantification and deeply limited the ability of pilots to improvise. And even with restrictions, the strategy was absurdly complicated. The actual application in warfare would be ridiculous. Indeed, that's exactly what the Navy brass concluded: Bomber-fighter strategies from RAND researchers were examined by naval research labs, but didn't make it far up the chain of command.[16]

Nonetheless, this work created a new line of mathematical research: *duel theory*, also called "theory of games of timing."[17] These timing games are completely artificial. They assume players act independently. They assume that past and future success are independent. They assume that players can strategize in terms of probability distributions of real numbers. But even with these restrictions, the mathematics is quite challenging, and many problems remain open. Though they would tell their bosses and military funders they were helping Cold War efforts, the mathematicians were just busy creating new puzzles for themselves to solve. Mathematicians have a particular affection for games of mathematics, and no matter where you employ them, they are going to look for the fun and challenge of new games.

―――

By the early 1950s, the game theory group realized that the program outlined by von Neumann and Morgenstern to quantitatively understand human behavior probably wasn't going to work without significant modifications. The clearest issue was that zero-sum games didn't really exist outside of play and sport. Even war was not really zero-sum. Expanding our understanding of game theory to non-zero-sum games was an immediate first step, and would later be responsible for John Nash's Nobel Prize.

But even in zero-sum games, there was little evidence that people played optimal strategies. This is still true today, despite a broader cultural fascination with game theory. Believe it or not, there are Rock, Paper, Scissors tournaments.[18] No one there plays the optimal strategy. In fact, in these tournaments, players "attempt to outwit their opponents through announcements, hints, taunts, or costumes."[19] Because the game is repeated, and because a player thinks they can outsmart the other person by repeated play, no one plays the optimal strategy. Merrill Flood immediately observed this problem at RAND in 1952. In human-subject experiments, Flood found that humans didn't play game theoretic strategies in the Prisoner's Dilemma. The Prisoner's Dilemma is the most culturally resonant artifact of game theory. The story of the game goes like this. Two prisoners are in a jail. Each prisoner is given one of two options: Either testify against the other prisoner or don't testify. If they both testify, they both get five-year sentences. If prisoner 1 testifies while prisoner 2 remains silent, prisoner 1 is set free while prisoner 2 gets twenty years in jail. If they both stay silent, both prisoners serve two years in jail. This game is not zero-sum, as the sums of the utilities associated with the different strategies are different.

Because of the setup of this game, the game theory optimal strategy is to testify. Think about it from the perspective of prisoner 1: If prisoner 2 testifies, then testifying is prisoner 1's best option, as it results in a sentence of five years rather than twenty. If prisoner 2 stays silent, prisoner 1 should still testify, and get out of jail rather than serve a two-year sentence. Switching the roles of the players says that prisoner 2 should also always testify against prisoner 1.

Flood and his colleague Melvin Dresher ran experiments on Prisoner's Dilemma, having their RAND colleagues as the players. They observed that after repeated plays, the player's strategies shifted. And no matter what, they didn't play a mathematically rational optimal strategy.[20]

Shockingly, Flood and Dresher observed that people didn't play optimal strategies in Prisoner's Dilemma before Albert Tucker had even posed the problem in the way we know it today. In Flood and Dresher's experiments, the game was presented solely in terms of a matrix of values to Flood and Dresher's experimental subjects. The captivating story of the two condemned prisoners scheming against each other played no role in the study.

Tucker invented the compelling story of the game only after seeing Flood and Dresher's experiment. It was certainly compelling and is widely used as a colloquial metaphor to describe real-life stalemates. But it does not at all describe how people resolve them! We knew the game-theoretically optimal strategy in Prisoner's Dilemma was suspect before it became a cultural meme. The lack of experimental reliability of game theory was apparent almost from the get-go. In the 1970s, this would be called the "folk theorem" of game theory: Two players in repeated games could end up with almost any set of payoffs imaginable.

In some ways, von Neumann and Morgenstern knew from the start that game theory was an unreliable predictor of human behavior. They note early on in *Theory of Games* that in Roulette, the mathematical expectation of the players is clearly negative. Thus the motives for participating in that game cannot be understood if one identifies the monetary return with utility.[21]

Perhaps even more concerning, even in single-play games people's decisions were strongly influenced by the phrasing of the game and their cultural backgrounds. This is nicely illustrated in decades of research on the game of Ultimatum.[22] In Ultimatum, player 1 is given 100 coins. They then offer some subset of these coins to player 2. If player 2 accepts the offer, then both players get to keep the coins: Player 2 receives player 1's offer, and player 1 receives the remainder. If Player 2 rejects the offer, both players receive nothing.

The mathematically rational strategy for this game is ridiculous. Player 1 offers player 2 one coin, and player 2 accepts. This is the optimal strategy for player 1, who would certainly walk away with a lower reward if they offered more coins. It is optimal for player 2 because something is better than nothing.

It's hard to fathom any instance of reality where two people would actually play this strategy. And experiments bear this out. But what's odd is the myriad of solutions people find depending on how the game is contextually presented. Sometimes player 1 offers half the coins. Sometimes player 2 accepts when offered ten coins. But the context is critically important, and it's hard to find any generalizable behavior. Imagine what happens if the amount of money changes: Players' behavior will be different if each coin is worth one dollar than if each coin is worth one thousand dollars. And imagine if each coin is worth one million dollars. Who could turn down an offer of a million dollars?

When the rules were rigid and explained clearly, it turned out humans didn't play optimally according to the axioms of game theory. Humans learned from multiple interactions. And outside the rigid confines of games, they might not even have "strategies" behind their decision making. Whereas the optimization models of chapter 2 were able to find niches where they were reasonable approximations of reality, game theory didn't describe how people actually acted *at all*. This work would inspire a controversial inquiry into how humans actually act, and that will be the topic of chapter 6.

But the game metaphor shaped the culture at RAND. As historian Robert Leonard summarizes it:[23]

> Repeatedly, one encounters the opinion that, as far as systems analysis was concerned, the value of game theory at RAND was not in providing numerical solutions to particular strategic problems, but in providing qualitative insight, in stressing the importance of reciprocation, opposing intentions, the credibility of threats, and the like. Game theory played a role in systems analysis, therefore, but not as a generator of numerical solutions to well-specified problems.

The metaphor of a game seemed like something promising to run with, even if the axiomatic game theory didn't lead to many fruitful applications. It is precisely in this loosely metaphorical way that game theory remains relevant in our current culture. And though it wasn't at all successful in quantifying our economic behavior, game theory would continue to have considerable impact on our understanding of actual games.

Just as the game theorists had realized that game theory was limited in its ability to predict economic behavior, computational gameplay hit a wall in the 1960s. The prototypes of Bernstein and Samuel were impressive, but both called for far more computing power than what was accessible. Computers were rapidly becoming faster, but to solve complex games like chess, they would need to achieve speeds unfathomable in the 1960s.

Bernstein's chess player ran on an IBM 704. The 704 could evaluate 3,000 chess moves in 8 minutes. Imagine it was 30 million times faster. Then it could evaluate a billion moves in 5 seconds. Such a hypothetical computer could evaluate all possible scenarios of 3 moves per player in less than 10 minutes. But technology 30 million times better than status quo seemed like a dream of science fiction. Despite the prescient optimism of Gordon Moore, most deemed it wholly implausible that computers could continue on their exponential trajectory forever.

Some AI researchers remained sanguine. John McCarthy, the researcher most people credit with coining the term "artificial intelligence," was convinced breakthroughs were on the horizon. In 1968, McCarthy bet David Levy, who was both a computer scientist and a master chess player, that in ten years a computer would beat Levy at chess. In 1978, when it came time to pay out the bet, Levy handily defeated the computer.[24]

But as we know now, the unfathomable happened. We now carry computers more than 30 million times faster than the 704 in our pockets. The IBM 704 is less powerful than today's run-of-the-mill microwave. The technological advancement between 1959 and 1997 was unprecedented. We sent men to the Moon, scaled up the infrastructure for aviation and transportation, revolutionized the production and reproduction of mass media, and achieved the massive speed-up in computers needed to make computer game players proficient. With transistor density doubling every two years, it was only a matter of time until not only Levy, but all human players were defeated.

That time came in 1997. In a celebrated match, IBM's chess-playing computer Deep Blue defeated grandmaster Garry Kasparov.[25] What did Deep Blue do? It ran Shannon's algorithm: It searched through game trees with modestly sophisticated ways of evaluating boards. On a supercomputer custom designed to solve these chess algorithms, IBM achieved the 1960s science fiction dream, an artificially intelligent agent that could evaluate 1 billion moves in 5 minutes. This required a machine that could perform 10 *billion* floating point operations every *second*. Shannon's program ended up playing out perfectly for chess; we just needed to wait fifty years for computers to catch up.

Since the 1990s, computers have only gotten more powerful. In 1997, Deep Blue was one of the most powerful supercomputers on the planet. In 2023, a PlayStation 5, a gaming console that one can buy in 2025 for $500, could compute 10 *trillion* mathematical operations per second. That's a thousand times the computing power of Deep Blue. And, of course, we use it to play games.

What about checkers? Checkers turned out to be simple enough that researchers were eventually able to compute the exact optimal strategy. Using hundreds of computers across 27 years, a collaboration of scientists enumerated game trees for checkers and determined in 2007 the best move for any possible layout of pieces on the board. It turned out that just like in tic-tac-toe, playing the optimal checkers strategy always leads to a draw.[26] Checkers would be the first of many games to be conquered by code.

In 1977, Hans Berliner devised a program to play backgammon that was able to beat a backgammon master in a few rounds.[27] Backgammon is different from chess and checkers because chance is a central component. Though Berliner's program was competent enough to win occasionally, it was also not sophisticated enough to consistently outperform skilled players. Fifteen years later, Gerald Tesauro unveiled TD-gammon, a backgammon playing program based on temporal difference learning (TD learning for short). TD learning was an algorithmically rigorous version of Samuel's machine learning scheme.[28] By 1995, it was clear that Tesauro's code was better than any backgammon player.

Even more fanfare came when computers conquered the game of Go. At first glance, Go appears considerably more complicated than Chess. The Go board is 19 × 19, and there are over 10^{170} valid Go boards as compared to around 10^{43} for chess. This is a monstrously large space. However, just like with chess, "conquering Go" did not mean finding an optimal strategy. Such an optimal strategy is completely out of the question for such a complex game. But since the space is so large, it should be clear that *humans* can't search through spaces so large in their heads either. The number of possible moves doesn't tell us how hard a game is, and a program might not need to enumerate all the moves to beat the best human Go players.

I've already discussed the main components of the winning Go system: tree search, value function approximation, and self-play. Each component needed innovations to build a winning system. Since the search tree of Go was so large, full enumeration of the search could never reach the depths possible in chess. A breakthrough occurred when gameplay researchers designed a clever approximate tree search algorithm to dig deeper into Go's game trees. Since Go was a game of perfect information, a deterministic strategy (which game theorists call a "pure strategy") would be optimal, but a strategy that deployed randomness might be still be useful in this. If the computer tried a random branch, perhaps it would stumble upon a good move. If it tried enough random branching, it could average its experiences and confirm that the promising moves would improve its current strategy. Randomness here helps speed up the searching through the near-infinite cascade of game trees. This random searching of game trees is a fundamental part of the algorithm Monte Carlo tree search (MCTS).[29]

Invented by Levente Kocsis and Csaba Szepesvári in 2006, MCTS proved to be incredibly powerful. By 2012, MCTS has mastered the 9 × 9 game of Go. And for the full 19 × 19 boards, it was ranked as a "4-dan" amateur, putting it in the top 5 percent of all ranked Go players.[30] The best Go players were ranked using this "dan" system. At 4-dan, players were very good. The best players were 9-dan. MTCS set us on the way to defeating the best human players, but it would take a second innovation to get there.

The second ingredient was a novel approach to approximate value functions. Just like in chess, from any board, the outcome of a game of Go under optimal strategies is predetermined as a win, loss, or draw. The key is being able to approximate this value even though we can't evaluate the quality of the board. Around 2014, researchers realized that they could approximate the value of a board by looking at past games.[31] Taking a large catalog of previously played games, they created a giant set of pairs of boards and outcomes. They then fed this dataset into a pattern recognition algorithm that predicted the outcome from the board alone using neural networks. Much to their surprise, they found that their machine learning models could predict the outcomes as well as highly skilled amateur Go players. I'll say more about why such pattern recognition was successful in future chapters. For now, the important point is that they used past games to build an estimate of the current value of any board. Combining these approximate value functions with

MCTS and mature temporal difference learning techniques, AlphaGo was soon able to beat any human player of Go.[32]

Of course, AlphaGo was assisted by massive computational power. In 2016, a team at the startup company DeepMind, with assistance from Google, built a program and assembled the necessary hardware to defeat master Go players. For the famous match against Go master Lee Sedol, the DeepMind team used a computer that could compute one million billion mathematical operations per second. This was 100,000 times more computing than Deep Blue. Moreover, to learn the approximate value functions, AlphaGo ingested information about 160,000 Go games, composed of around 100 million moves.

I want to reiterate again that AlphaGo did not find an *optimal* Go strategy and Deep Blue didn't have an *optimal* chess strategy. Leaning on seminal ideas from the 1950s and seven decades of exponential growth in computing, both found strategies that game experts are convinced will beat all humans. As much as it pains us to admit it, bettering humans is considerably easier than finding the absolute optimal strategies. Because of the daunting complexities of chess and Go, we'll never find an optimal solution for them without a complete intellectual revolution.

Perhaps from this narrow perspective, von Neumann and Morgenstern were right. Chess *was* trivial once you had enough compute. But what about *"interesting"* games of imperfect information? These games would prove harder to crack, but with a few key insights, even these soon fell to the computers.

Early in her PhD, Daphne Koller thought game theory might provide keys to understanding artificial intelligence. Whereas

Samuel and Shannon focused on machines as single adversaries, Koller was interested in understanding how multiple machines might interact and make decisions together. Her master's work was on *consensus* in distributed systems. Which rules do you need to make to ensure that multiple computers could coordinate to complete a task and the task would still be completed even if some of the computers failed?

In her PhD, Koller was interested in a broader set of strategies. She started reading about game theory as a potential general theory of coordinating distributed agents. By the time Koller was working on game theory, most behavioral and social scientists accepted that game theory couldn't help with understanding how humans made decisions. But Koller was interested in whether it was a useful formal model for *computers* to make decisions.

Koller quickly realized that even for problems that seemed simple, the computational tools for games with multiple players and uncertain information were completely unimplementable. The first reason was that the game-tree techniques couldn't be directly applied. As I mentioned earlier in this chapter, players hide information from their opponents in imperfect information games like poker. Opponents don't show each other their cards. This is in stark contrast to games like chess or Go, where all of the game information is out in the open on the board.

But in the early 1990s, the only alternatives to game trees were tables of strategies. And for most games, these tables were impossible to compile. To explain why, let's focus on games with only two players. The *normal form* of a game writes a game as a table like the ones we drew earlier in the chapter. Each row of the table corresponds to a deterministic strategy for player 1. Each column of the table corresponds to a deterministic strategy for player 2. The entry in row *i*, column *j* specifies the reward

TABLE 4. Simplified outcome table for Rock, Paper, Scissors

0	−1	1
1	0	−1
−1	1	0

player 1 receives if player 1 plays strategy *i* while player 2 plays strategy *j*. Table 4 shows a compact way to rewrite the table for Rock, Paper, Scissors.

For games with only a few strategies, it's easy to write down these tables and philosophize about human psychology. Games of Prisoner's Dilemma and Ultimatum are both represented by 2 × 2 tables.

But what is the size of a table for a game like poker? Think about even a simple variant of the game. Here's a dramatically simplified version of poker invented by Harold Kuhn:[33]

> There is a deck of three cards, a Jack, a Queen, or a King. The game proceeds with each player anteing 1. Then each player is dealt one card from the deck. The third card is not used in the game, and no one gets to see what it is. The game starts with player 1 who can bet or check.
>   1. If player 1 checks, then player 2 can check or bet 1.
>   1a. If player 2 checks, the higher card wins the pot
>   1b. If player 2 bets, player 1 can fold or call.
>   1ba. If player 1 folds, player 2 takes the pot.
>   1bb. If player 1 calls, the higher card wins the pot.
>   2. If player 1 bets, then player 2 can fold or call.
>   2a. If player 2 folds, player 1 takes the pot.
>   2b. If player 2 calls, the higher hidden card wins the pot.

In Kuhn's poker, there are only 3 cards, not 52. There is one round of betting. And already things are far more complex than in Prisoner's Dilemma. What does a pure strategy look like for player 1? They need to specify two decisions: First, what will they do when dealt a particular card, and second, what will they do if they check and then player 2 bets? Each decision point has eight strategies (specifying check or bet based on which card they have in their hand). Hence the total number of strategies for player 1 is 64. Player 2 also has 64. So for the most simplified version of poker conceivable, the table approach to solving the game is too large to inspect by eye. If we went from 3 cards to 10, the table for Kuhn poker would have over a million rows and columns. Can you even imagine what it might be for Texas Hold'em?

Koller was puzzled by the computational intractability of imperfect information. Surely someone must have thought about simplifying this tabular representation of imperfect information games.[34] Her PhD Advisor connected her with Nimrod Megiddo at IBM Almaden. Koller expected Megiddo to have a simple solution, but he instead thought her problem was quite tricky and worth studying.

Together, Koller and Megiddo found an astonishingly smaller representation. Their key breakthrough was to derive strategies from sequences. The state of a game like poker is defined by the sequence of events that have occurred thus far in a game and are observed by a player. For example, in Kuhn poker, suppose player 1 is dealt a jack. Then the sequences are the first move sequences—either bet or check—and the second move sequences when player 1 checks and player 2 bets: (check, fold), (check, call). Since there are 3 cards, player 1 has $3 \times 4 = 12$ sequences. If you do a similar counting, you'll find player 2 also has 12 sequences. If there were 10 cards in this case, each player would have 40 sequences. The massive blow-up of options has stopped because we are enumerating sequences instead of strategies.

The number of sequences in a game is the same size as the number of nodes in a game tree. This is because every node in the game tree comes from a unique sequence of events in a game. The number of strategies for any realistic game, on the other hand, is effectively infinite.[35] Koller and Megiddo's realization was that a game could be formulated entirely from consistency on sequences. Suppose players were playing mixed strategies, so their choices were random. Then, it must be true that the probability of arriving at any particular game state had to equal the probability of entering any of the states accessible from that point. That is, I could only get to the next part of a sequence if I had already gotten into the first part of the sequence. With Bernhard von Stengel, they showed that the consistency rules resulted in a linear programming formulation for the game that was exponentially smaller than the normal form that had been standard since the forties.[36] And this made solving poker begin to look possible.

After her PhD, Koller went to Berkeley to do a postdoc. Having nailed down a theory for sequence form games, it seemed like she should be able to solve real imperfect information games. She wanted to write the software. With undergrad researcher Avi Pfeiffer, she built a system that was able to solve the first games resembling poker.[37] As had been expected but not proven before, Koller and Pfeiffer's code proved definitively that bluffing was required for an optimal strategy in poker. Pfeiffer went on to make a general language for specifying and solving imperfect information games, but most people just wanted to keep studying poker.

Though Koller's work on sequence form games was an academic home run, impressing researchers in artificial intelligence, economics, and computer science and helping to secure her a professorship at Stanford, Koller stopped working on games at this point. She was dissatisfied with the limitations of game theory. Game theory was a beautiful mathematical model, and it could clearly solve deep math problems. However, she determined these problems would only lead to solutions to . . . games. Koller wanted to work on problems that would actually make a difference to society and people, and she decided game theory couldn't help.

"Understanding the world around us is more important than understanding the optimal way to bluff," she told me. In her experience, when she needed to model people in simulations of complex systems, modeling their decisions as random got her 90 percent of the way to a solution. How to best make decisions under wide-ranging uncertainty was far less cut-and-dried. For Koller, once you stepped away from the game board and had to make decisions in reality, understanding uncertainty and the myriad ways it could arise and impact plans was more important than strategy.

Koller, Megiddo, and von Stengel's sequence form caused a stir in the world of poker bots. Before the sequence form, computer poker tried to emulate the tree search methods of chess, using a lot of statistical simulations to make up for the fact that there was imperfect information. But sequence form provided another route.

Led by Jonathan Schaeffer (the same person who led the team that solved checkers), a team at the University of Alberta set out to build a poker-playing robot based on sequence form. The simplest "real" poker game is two-player Limit Texas Hold'em. This game is similar to the game played in the World Series of Poker except the sizes of the bets are limited (so you can't go "all in"), and each round has at most four bets. You can play Limit Hold'em at many casinos on the Las Vegas Strip. Even this simple form of poker has a game tree with 10^{18} states. This tree is smaller than that of checkers, but was too large to solve using linear programming in the late 1990s. So the team used *abstractions* of poker, clustering hands together as effectively equivalent, and dramatically reducing the number of sequences they had to consider.

Abstraction makes the search space smaller by forgetting certain details. Some obvious hands can be bucketed together. In the first round of Hold'em, players are dealt two cards face down and then they bet. At this stage, any hands with two queens are equivalent to each other. There are six such hands: club-diamond, club-heart, club-spade, diamond-heart, diamond-spade, heart-spade. Grouping these immediately reduces the size of the game. But hands that are not quite equivalent but have approximately the same "hand strength" could also be lumped together. With enough abstraction,[38] Hold'em could be reduced from 10^{18} states to 10^{7} states. This was now of the size that could be solved in less than a day by the linear programming software that was readily available in the early aughts. And the strategies returned by the linear program easily outperformed the existing bots.[39] The team was clearly on the right path.

At this point, Schaeffer had taken on an administrative position at Alberta, and asked Mike Bowling to lead the poker efforts in his absence. Bowling's team set out to find better abstractions. However, the linear programming computations were slowing the process down. If it took a day to solve the sequence form game for every small change in abstraction, it would take forever to explore all the possibilities. They either needed faster computers or better algorithms. They decided to look for the latter.

Bowling's friend Martin Zinkevich had followed him from Carnegie Mellon University to Alberta and joined the lab as a postdoc. Zinkevich thought he could speed up the solution of sequence form problems by parallelizing over multiple machines. On his way to a parallel solution, however, he invented a completely different algorithm, now called *counterfactual regret minimization* (CFR).[40] As is now a recurring theme, CFR is based on self-play: It effectively has a poker bot play itself and updates its strategies with successive information. And instead of the day-long solve of the linear programming approach, CFR could solve the abstracted poker problems in an hour. Since it was now so fast, Bowling realized that maybe instead of looking for better abstractions, they could use CFR to actually solve poker.

With some other algorithmic insights and a lot of computer engineering, Bowling succeeded in solving the first "real" version of poker. His team figured out how to compress the probabilities over 10^{17} states in 10^{13} bytes. This compressed file was analyzed by 200 compute nodes for days. They figured out how to maximally utilize the computers, running them to such extremes that they had frequent hard drive failures because the drives were never spun down. But in the end, they could declare that two-player Limit Texas Hold'em was solved. They had a developed a computer program that, if played against for long enough, would beat anyone.[41]

Computers conquering poker didn't end poker. Instead, it made it possible for every human player to be better. Intrepid coders saw these breakthroughs in game solvers, wrote up their own implementations, and then sold them to poker players. Commercially available packages run CFR to solve poker from restricted positions and give players the expected values of winning. Professional poker players, like those seen on the World Series of Poker, spend hours a day staring at the output of computer solvers and memorizing moves with higher chances to win.[42]

Even without perfect solutions, raising the floor for humans has occurred in every instance where computers became better at games than people. For example, professional chess players train themselves using computer chess engines. In the big book of games they practice with every day, they've added "Alpha-Zero" games, generated by a chess engine that uses the same algorithms as AlphaGo. These computer moves are so distinctive that players identify certain chess moves as "Alpha" inspired.

In some sense, it's not surprising that superhuman solvers didn't kill gameplay. Someone whose hobby is to play basketball doesn't stop because they aren't in the NBA. They find people at their skill level and play them. And they still strive to be the best they can. Superhuman computer solvers made people better at the games! As Bowling observes, "In all of these games where there was a vibrant computer community, the skill level of the human community took a massive jump."

Though the language of game theory still occupies plenty of discourse about human interaction, as a means to actually understanding humans and human economies game theory was a failure. The complexities of social and economic interaction can't be shoehorned into rigid, rule-based games; the complexity of computing solutions to even highly simplified models of reality has proven intractable; and, most importantly, humans are decidedly not mathematically rational. Game theory is useless at predicting how humans will behave when complex decisions are laid out before them. There is no fix for this. As I'll describe in more detail in chapter 6, diversions into understanding how humans are "predictably irrational" have been sullied by deep and fundamental replication crises and high-profile fraud scandals.[43]

But that doesn't mean game theory was useless.

Computational gameplay has pushed computers forward, captivating engineers and people to build machines with better capabilities. And it has given us insights into how we play games. Playing games is part of being human, and these sorts of competitions help connect us with other people and challenge ourselves to improve. It's fine that what turned out to be interesting about computer game play was what it taught us about the games themselves, not what it taught us about some pie-in-the sky conception of artificial intelligence.

What did computers teach us about chess? Shannon posited that you could build a chess-playing robot with efficient tree search and reasonable board evaluation. This turned out to be correct, and proving it correct was simply a matter of waiting fifty years for engineers to scale up computers. What did computers teach us about checkers? Checkers is indeed so simple that exhaustive search can and does find the optimal strategy. Playing optimally in checkers ends in a draw, just like in tic-tac-toe. What did computers teach us about Go? Computer science researchers in the 2010s realized that a simple "code book" that just looked at the board and guessed the next move could probably beat amateur players. Even though Go *looked* harder than chess because of its massive game trees, a computer didn't require impossibly deep tree search to beat people. Maybe in retrospect this is not that surprising, and, in fact, perhaps Go is actually in many ways *easier* than chess? What happened in poker? Around 2010, researchers realized you could find optimal betting strategies for small subgames of poker by writing the game in the correct way. This led to the solving of Limit

Hold'em and competitive play at No Limit. And then professional poker players started memorizing expected value tables from poker solvers so they could play "game theory optimal" in big tournaments.

In all these cases, computers taught people how to play better. Did our pursuits into computational game theory teach us anything about building superintelligence? No. Did it give humans competitive advantages so they could go best other humans in parlor games of skill and chance? Yes.

A second important contribution of game theory was providing a particular means of participatory decision making. It gave people a language to frame thorny, complex decisions as games. Game theory works when it is a *prescriptive* model of behavior, not a descriptive one. Game Theory provides a programmatic, though imperfect, solution framework for complex problems with many involved parties with competing interests.

Stakeholders first have to agree that *a mathematically rational solution is an acceptable solution.* In many tricky problems, it is impossible to make everyone happy with a cut-and-dried solution. By accepting the decisions of a rational algorithm, they at least defer the agency to a third party with clear, transparent rules for deliberation. The people affected by the decision can debate and agree upon the acceptable utilities associated with each outcome. And then the policymaker can design a game-theory optimal solution to the agreed-upon formulation of the problem. People need not be mathematically rational, but they can agree to accept a mathematically rational decision.

One of the most famous success stories of this sort of decision making is the National Resident Matching Program (NRMP), which matches graduating medical students to residency programs across the United States. Designed in the 1950s, the matching program has residents list their preferred programs and hospitals list their preferred candidates. Collecting these lists of preferences in a national clearinghouse, it applies an algorithmic procedure to find a suitable assignment for all involved parties. Deferring to an algorithm eliminated a haphazard, unfair, and chaotic process.[44]

Though it had not been explicitly designed using game theory, RAND Alumni David Gale and Lloyd Shapley determined in 1962 that the NRMP was game-theoretically optimal insofar as no party could get a better outcome for themselves if they didn't tell the truth about their preferences.[45] However, this coincidental game theoretic optimality was only part of why the NRMP has been a marked success. Because it had transparent rules, when the NRMP showed flaws, its rules could be patched. Notably, the. NRMP adjusted its policies to accommodate married couples being matched to nearby hospitals. The transparency of the mechanism and the fact that it had well-specified rules made it easier to figure out how to fix the deficiencies that inevitably arose through its implementation.

Organizations have deployed the same sort of game theoretic mechanism design to allocate communication channels for telecom providers, to assign children to high schools, and to find donors for kidneys. These solutions do not make everyone happy, and they need to be revisited as unforeseen impacts are discovered after the implementation. But game theory does provide an initial step in certain cases to provide an acceptable outcome to many challenging decision problems. It's not the only game in town, but it can be a valuable tool in a mediator's toolbox.

4

Regulations, Regularities, and RCTs

ALEXANDER FLEMING discovered the antibacterial properties of penicillin in a petri dish in 1928. This discovery would transform medicine, curing formerly fatal infections. But how did we figure out how to treat humans with this antibiotic?

In 1941, doctors in the UK, led by H. W. Florey, started treating bacterial infections with penicillin. The results were dramatic. Patients who would likely have died otherwise showed remarkable, rapid improvement after penicillin administration. Florey's team reported their experiences in the 1941 *Lancet* article "Further Observations on Penicillin."[1] Their "Therapeutic Trial of Penicillin" consists of ten case studies. Yes, that's right, only ten patients. But these case studies are outlined in gripping detail.

In the first case, a 43-year-old police officer had a severe staph infection. The infection had spread from the corner of his mouth to his eyes and down his right arm, infecting the bone. On February 12, they gave him a dose of penicillin. After the first day, they already noted improvement in his condition. But by February 14, they thought his condition was not improving and upped his dosage. In 1941, penicillin was still rare and hard to manufacture. The doctors had to be careful to use as little as they could get away with. To raise this patient's dosage, they had to recover penicillin from his bedpan. But these extreme measures proved fruitful, and by February 16, the patient had markedly improved. All of his infections were swelling less, and some of the regions looked almost back to normal. The patient felt better, had a full appetite, and had a normal temperature. The hospital ran out of antibiotics on February 17, but the patient's condition remained temporarily stable. It was clear that he still had an infection, but it didn't seem to be getting worse. Sadly, the patient's condition deteriorated ten days later. With no penicillin left, the team could only watch the infection run its course. On March 15, the patient died.

The doctors concluded that antibiotics could not be stopped early. The treatment could only be effective if they gave a large enough dose. In a second case, a 48-year-old laborer presented with a worrying bacterial infection in his shoulder. They kept this patient on antibiotics for eight days, a day after the infection seemed resolved. The patient recovered more or less without incident.

The final four cases of the paper discuss the topical use of penicillin to treat eye infections. All four of these cases were more minor infections, but all patients rapidly recovered from relatively small doses of the antibiotic.

From these ten patients, six with severe bacterial infections and four with eye infections, the doctors drew many conclusions. First, unlike other antibiotics of the day, penicillin seemed not to be toxic to humans, even at adult doses of up to 1g a day. Second, even though some of the patients had died in their trial, penicillin prevented the worsening of bacterial infections in every case. Every patient's temperature fell and their infections all locally improved. Third, penicillin was able to cure the infection in several cases. Finally, in the six serious case studies, all but the laborer had received other drugs. Some had even had surgery to attempt to purge their infections. None of these treatments had proven successful. But penicillin changed the course of the illness in every case. The results were dramatic. Patients who would likely have died otherwise showed remarkable, rapid improvement after penicillin administration.

All of these observations turned out to generalize well beyond this study. Penicillin was rapidly manufactured in the United States to help prevent infections of soldiers wounded in the imminent invasion of Normandy in World War II.[2]

From our modern view, the idea that you could go from a few cases in a single hospital to mass manufacturing for a war effort seems unimaginable. The bureaucratic regulations on therapeutics are thick, and the number of successfully treated patients required to satisfy regulators is vast. But why are today's requirements so stringent? How did clinical medicine go from a field that relied heavily on observing what was obvious to one that required vast statistical recordkeeping?

Therapeutics that "work" are a relatively new innovation in medicine. The 1930s and 1940s brought not only antibiotics but countless pharmaceutical interventions. Insulin was discovered in 1921, and mass production to treat diabetes began in 1923.[3] The 1930s saw widespread manufacturing and popularization of barbiturates.[4] A few years before the initial human studies on penicillin, sulfonamides were widely developed to treat bacterial infections and deployed in the Allied war effort. Antihistamines like Benadryl were first applied in the clinic in the 1940s.[5] Steroids for treating rheumatoid arthritis were first clinically applied in 1948.[6] Innovations in chemotherapy came in the 1940s, notably with the discovery of aminopterin to treat leukemia in children by Sidney Farber in 1948.[7]

Just as in the case of penicillin, all these treatments "worked" with far more efficacy than their nineteenth-century counterparts. They were not without controversy. Sulfonamides, in particular, were plagued by myriad adverse reactions. Over 100 people across the United States died from poisoning by a bad batch of sulfanilamide. The drug was suspended in antifreeze, and no one had bothered to test whether this liquid suspension was toxic.[8]

In response to the sulfanilamide disaster, Congress passed the Federal Food, Drug, and Cosmetic Act in 1938. As part of this bill, congress granted the FDA powers to oversee this new zoo of pharmaceuticals. If drugs were going to be made widely available, the public needed to be assured that they were effective, reliable, and safe.

If not all drugs were curative, and if some drugs had rare adverse events, how could we find them? The revolutionary idea was applying the statistical methods of *biometry*. The impact of drugs would be assessed through careful experimentation that removed biases of practitioners and collected statistics of harm and benefit. The proposed way forward was drug evaluation through rigorous *randomized clinical trials*.

Though not explicitly computational, the randomized clinical trial was driven by the parallel growth in statistical bookkeeping and tabulation during the postwar period. In our present, the randomized trial has become a favored tool for unbiased experimentation both inside and outside of the medical clinic, applied widely in development economics and in large-scale design and feature testing by technology companies. The randomized trial is often referred to as a "gold standard" for causal decision making,[9] providing a mathematically rational path to find whether an intervention works or not. The promise of the RCT is that we can make rational decisions simply by randomizing and counting. In this chapter, I will hone in on exactly what we can and cannot learn from randomized clinical trials. Establishing that an intervention "works" will turn out to be fraught with complexity.

In 1946, statistician Bradford Hill helped design the first randomized clinical trial to test whether streptomycin, yet another antibiotic, was effective in treating tuberculosis. Tuberculous had proven much more challenging to treat than the staph infections seen by Florey's team. Penicillin itself was ineffective in curing tuberculosis. But in the mid-1940s, Selman Waksman and Albert Schatz discovered streptomycin in soil and found it killed tuberculosis in guinea pigs. Efforts were quickly made to test it in people.

Though not a physician himself, Hill's statistical work almost exclusively focused on medicine and epidemiology. He had dedicated his life to developing mathematical systems to organize and interpret medical data, and had been working on the design of medical trials for a decade. He was keen to apply his novel experimental designs. Hill's innovation was to borrow ideas popular in agronomy, enrolling patients *at random* into treatment and control. This randomization removed several sources of bias. In past trials, physicians could tip the scales—consciously or unconsciously—by manipulating who was allowed into a trial and who received the new therapeutic. If physicians were free to choose which people got a drug and which didn't, perhaps they would only give the drug to the less frail people to make the drug look a bit better. The selection of who went into treatment or control was best taken out of the hands of the trial physicians. Prior schemes to avoid such favorable selection biases relied on alternation, where patients were enrolled in treatment and control based on the order in which they entered the trial. But a clever doctor passionate about the treatment could still easily deduce which patients were getting the drug. By randomizing and blinding the assignments, it would be impossible for these biases to creep into the trial design.

In the streptomycin trial, each hospital was given envelopes containing the letters "C" or "T." These envelopes were shuffled into two piles, one for men and one for women. When a new patient was admitted to the trial, the team opened an envelope in their corresponding pile and gave the treatment associated with the enclosed letter. Though the doctors were not perfectly blinded, the patients themselves were not told that they were in the trial.[10]

In the trial, the streptomycin appeared to be superior to bed rest alone. Only 4 of the 55 given streptomycin died. On the other hand, 15 of the 52 assigned only to bed rest perished. These numbers give us an opportunity to think about how to analyze the results of the trial. Around 25 percent of the patients with tuberculosis died when only prescribed bed rest. On the other hand, over 90 percent of the patients treated with streptomycin survived. Since the only difference between these two groups was the administration of streptomycin, perhaps we would infer a similar benefit for new patients. One interpretation would be that we should expect a fourfold reduction in deaths in our hospital.

A subtle conceptual shift has happened in the move from penicillin to streptomycin. This shift is now so pervasive that we seldom notice unless someone (like me) belabors the point. In the study of penicillin, the clinical reports discussed what happened to particular individuals and made arguments about what to expect in future patients based on these specifics. Understanding the implications of the streptomycin trial requires a different sort of reasoning. The argument for an effect is now in terms of percentages, and the implication is that such percentages will transfer to future experience. The results suggest that in the next tuberculosis season, we'd hopefully have four times fewer deaths in the tuberculosis ward, provided that streptomycin is widely prescribed. The trial does not indicate which patients will do better. Based on the experiment, we can't know who might respond. Though the individual patients will all likely differ, the hope in a randomized trial is that there is some similarity between *groups of patients* that lets us reason about percentages of outcomes at the group level. We move from conceptualizing individual outcomes to conceptualizing *rates* of outcomes.

Part of the reason that streptomycin needed to be analyzed in terms of outcome rates was that it was not obvious if it was useful for treating tuberculosis. The trial revealed that it was indeed not very effective. While it was much better than doing nothing, precisely how much better is up for debate. Four out of 55 is an *estimate* of how many people died based on the limits of a particular trial. We don't know that only four people would have died in the control group if they had been given streptomycin. We can only muse about the counterfactual scenario of the control group given our observation of the treatment group.

Statisticians deal with this counterfactual uncertainty using a thought experiment called a *null hypothesis*. In the imaginary world of the null hypothesis, we assume that streptomycin does nothing at all. Instead, we had a total pool of $55 + 52 = 107$ patients, and $4 + 15 = 19$ of them were destined to succumb to tuberculosis. All our experiment did was split this group in two at random. What would happen if we took this set of people and randomly flipped a coin, adding them to the treatment group if we got a head and the control group if we got a tail? In the thought experiment where the treatment did nothing, we'd still have seen a different number of deaths in the two groups. You can try this for yourself. If you flip 19 coins, you might see 8 heads, you might see 11 heads. But what are the odds we'd see at most 4 heads? The answer is roughly one hundred to one.[11] Belaboring the point to spell out the reasoning, under the null hypothesis in the streptomycin trial, the odds that we'd see at most four deaths in the treatment group assuming the streptomycin did nothing is roughly 100 to 1.

If the odds of a null hypothesis are better than 20 to 1, scientists have widely adopted the convention of saying that the trial result is "statistically significant." I don't want to dwell on this confusing notion in the chapter. We don't need statistical significance to understand the history of trials. "Statistical significance" is just a rubber stamp saying the odds that our data could have arisen from the funny thought experiment are low. It's nothing more than a regulatory bar to pass. We set up the rules of the game so that the minimum standard for future conversation is that the odds the measurement could occur if the treatment did nothing are more than 20 to 1. Is this the perfect rule for scientific understanding? No. The threshold of 20 is pretty arbitrary, and people tend to argue a lot about that threshold without looking at the other experimental details. But if we consider it merely a loose bar of acceptance, we're more likely to be satisfied with an imperfect rule. "Would you be surprised if you won when those were the odds?" The more surprised you are, the more you might believe the result.[12]

In the case of streptomycin, 100 to 1 odds seemed reasonably surprising. The fourfold risk reduction seemed promising but not definitive. In terms of the percentages of patients that could potentially be saved, streptomycin was better than nothing.

Unfortunately, further observation during the trial led to more disappointment. After six months, tuberculosis could still be cultured from 47 of the 55 patients in the treatment group versus 50 of the 52 in the control group. After antibiotic-resistant bacteria formed in the treatment group, the trial was thus ended early. An effective treatment for tuberculosis, consisting of a potent cocktail of different antibiotics that would be administered for months, would require a few more decades to develop.

Following the streptomycin trial, Hill gave a series of lectures to persuade the medical community of the value of the clinical trial. To do so, he needed to get them on board with a conceptual leap. Doctors had always focused on individual patients and treatments. Hill was arguing for thinking about the statistics of groups of patients. This was a big ask. Though statisticians had developed powerful mathematics to formalize probabilistic statements about the world, their terminology and techniques were confusing. Doctors would need a lot of convincing to bring statistics into their clinic. Statistics wasn't needed to understand that penicillin worked, but it helped clinicians understand the effectiveness of streptomycin. Let's look at Hill's arguments for what differentiated these cases and when statistics could benefit medicine.

Statisticians spent much of the early post-war period establishing ground rules for when their methodologies were applicable to societal systems. For example, M. S. Bartlett, a mathematical statistician, addressed the Manchester Statistical

Society in 1951 to clarify the needed starting points for a statistical calculation:

> Insofar as things, persons, are unique or ill-defined, statistics are meaningless and statisticians silenced; in so far as things are similar and definite . . . they can be counted and new statistical facts are born.[13]

Statistics was about counting, but only about counting things that were not unique and things that were well defined. Statistics had no business musing about the singular and the unverifiable. Bartlett further emphasized, "Our arithmetic is useless unless we are counting the right things." But what were "the right things" in the context of medicine? Hill tried to expand on this in his lectures on clinical trials.

> The most frequent and the most foolish criticism of the statistical approach in medicine is that human beings are too variable to allow of the contrasts inherent in a controlled trial of a remedy. In other words, each patient is "unique" and so there can be nothing for the statistician to count. But if this is true it has always seemed to me that the bottom falls out of the clinical approach as well as the statistical. If each patient is unique, how can a basis for treatment be found in the past observations of other patients? In fact, of course, physicians do not act like that. They base their "method of choice" upon what they have seen happen before—whether it be in only two or three cases or in a score. But even if human beings are not each unique in their responses to a given treatment they are certainly likely to be variable, sometimes extremely variable.[14]

If everyone is unique, then it is likely impossible for past practice to inform future treatment. There must be some regularity in disease progression, or healthcare would be impossible. But Hill also points out that if everyone is the same and responds the same way to a treatment, as seemed to be the case with penicillin, there's no need for statistics either.

Hill was threading a fine needle with this reasoning. Things that are easy to predict don't need statistics. Things that are singular or unpredictable can't be interrogated with statistics. It is in the very murky middle, where things are highly variable but potentially predictable, where statistics might help.

Predictability alone doesn't fully characterize the narrow scope of statistics. Statistics is only needed when there is uncertainty about counterfactual scenarios. When a particular outcome is impossible without the intervention, we again don't need statistics!

In the penicillin case studies, people who would never have recovered were returned to health. The counterfactual scenario, "That patient would have lived had they not been treated," was implausible to the attending physicians. In a similar vein, Sidney Farber published a radical new chemotherapy treatment of acute childhood leukemia in 1948. Before chemotherapy, no child had ever been observed to recover from leukemia. Chemotherapy resulting in remission in a single patient was almost enough to establish that the treatment caused remission, as the alternative is impossible. Hence, Farber's results with a few children were a total revelation.

On the other hand, Hill notes that there are diseases with far less clear-cut outcomes and treatments. He describes rheumatic fever, where "some may die, some may have prolonged illnesses but recover eventually with or without permanent damage, some may make immediate and indisputable recoveries." In such a case, the inherent variability of outcomes makes it more challenging to establish a causal connection between treatment and cure. Similarly, people often survived tuberculosis infections. If there were only case studies, we could plausibly imagine that an individual treated with antibiotics would have survived without the treatment.

Statistical reasoning, like so many other computational tools, has a sweet spot. If a treatment always works, there is no need for statistical counting. If an outcome is too variable or unpredictable, statistics can't help either. We can only apply statistics when a disease prognosis is bad but not definitively fatal and the associated treatment is good but not definitively effective. We end up using statistics to approve treatments that work for some but not for all, without any insights about for whom exactly these treatments will work.

But what's the use of such a tool in medical practice? How does this help doctors treat their patients? While Hill believed strongly in the value of the randomized experiment for such practice, he also had a response to those who remained skeptical. Perhaps by "counting the right things," "overall" health would improve *in the general population*. Statistics could be used to inform broad-reaching policy. This was the sweet spot, and the randomized trial would find its greatest success when deployed at a national scale in an attempt to fight polio.

―――

In preventive medicine, a *healthy* patient is given a drug or other treatment. The desired outcome is *not* contracting or succumbing to some specific infection or other pathology down the road. A measles vaccine should prevent measles. Regular sunscreen use should prevent skin cancer. Wearing a surgical mask should prevent the acquisition of respiratory illness. In all these cases, a person's potential outcomes are murky and uncertain.

Preventive treatments are necessarily harder to evaluate than curatives because we can never know in advance whether a person would have acquired a disease to begin with. For any particular person, it is impossible to ascertain whether the treatment prevented a disease or if they never would have acquired it anyway. People have wildly varied susceptibilities and resistances to various diseases.

Preventive medicine checked all the boxes necessary for statistical inference. Epidemiologists like Bradford Hill actively engaged in research to promote statistical methods in evaluating public health programs aimed at prevention.

Hill was particularly enamored of vaccine trials. He helped advise the first British randomized trial of vaccines. In his 1957 lectures to the Royal Institute of Public Health and Hygiene, Hill laid out why vaccines were so challenging to evaluate.[15] Just as with the rapid advancement of other therapeutics, the number of vaccine products exploded in the years following World War II. Hill exclaimed, "The dimpled arms and buttocks of the children of Western civilization are fast being reduced to pin cushions. This clearly has its disadvantages."

Hill may have been joking about children's association of doctor's visits with needles and pain, but he was seriously concerned with the potential harms of the vaccines themselves. He was well aware that vaccines could cause grave injuries, and the risk of these rare but serious harms had to be balanced against the potential benefit of any vaccination program. He concluded that the controlled randomized trial was the best method to assess these tradeoffs.

Hill's arguments for vaccine trials were again the mitigation of selection biases. He worried that volunteers for vaccines would be different from those who abstained. Perhaps these would be people who were more generally cautious about their health and might avoid infection by behavioral means unrelated to the vaccines. For example, they might avoid crowded places when infection rates were high. Moreover, for certain diseases like the flu, Hill worried that people wouldn't be keen to volunteer for vaccination for a disease where the odds of severe illness, or even illness at all, were low.

A placebo-controlled trial could eliminate these potential sources of bias. Even if the particular volunteer group was narrow, the effectiveness of a vaccine on this group could indicate effectiveness for a broader population. Moreover, vaccines are one of the easiest treatments to blind. Saline shots can be given as a placebo, and the identity of what a person was inoculated with can remain hidden until a final outcome is ascertained.

On the other hand, vaccine efficacy was considerably more difficult to adjudicate than therapeutic efficacy. Drugs are administered to sick people and judged a success if the sick convalesce. Vaccines are administered to the healthy and judged based on how many people eventually fall ill. In the streptomycin trial, the outcome to be judged was simply death. For vaccines, since almost no one in the trial would even get sick, only a much weaker condition could be evaluated. The endpoint might be "symptomatic illness," where a patient both displayed characteristic symptoms and had a laboratory-confirmed infection. These diagnostic criteria were subject to the (potentially biased) judgments of the trial staff. Moreover, for any individual in the trial, it would be unclear whether they stayed healthy because their vaccine worked or because they would never have contracted the disease in the first place. Though these issues seemed minor, they would immediately complicate interpretations of early trials.

Hill helped advise initial vaccine trials in the UK, including an extensive study of a vaccine against whooping cough.[16] The whooping cough trials were a tremendous undertaking, testing four different vaccines with nearly 10,000 participants. But this trial would soon be dwarfed by the most ambitious vaccine trial of all time.

In the 1940s and 1950s, polio was a terrifying disease that would paralyze young children. But polio's epidemiology was peculiar. Polio seemed to afflict those of higher socioeconomic status more severely, likely because poorer children were conveyed immunity from their mothers who had probably already been infected. On top of this, symptomatic polio infections were also very, very rare. During the 1940s and 50s, there was only around 1 case of paralytic polio for every 4,000 children in the United States. As a result, understanding treatment and prevention was a daunting challenge.

Jonas Salk developed a potentially lifesaving vaccine. He desperately wanted to mass produce and administer this treatment to prevent the ravages of polio from afflicting young people everywhere. But Salk's vaccine was itself a scary prospect. The vaccine was based on a live virus, and specialists worried that this might cause the disease itself. Before it could be used as a cure, it needed to be carefully tested.

A massive trial was designed to test the efficacy of the vaccine. Thomas Francis, a virologist who had experience with prior vaccination campaigns, was recruited to run the trial. Francis argued that removing bias was essential to establish public trust that the vaccine worked,[17] and, after consulting with Hill, only agreed to run the field trials if they were randomized and placebo controlled.[18]

Evaluating the safety and efficacy of such a novel and potentially risky vaccine was a challenge for such a rare disease. In order to rule out the null hypothesis that the vaccine did nothing, the trials would need to record about 100 cases of paralytic polio in the control arm of the trial. At a rate of 1 bad case of polio per 4,000 participants, vaccine trials would need hundreds of thousands of children.

This was indeed the scale at which the Salk vaccine was trialed in 1954. It remains one of the largest clinical trials ever performed. Francis met political headwinds in his advocacy for randomization. Since some states wouldn't agree to his randomized design, an alternative plan was offered where second graders were offered the vaccine and first and third graders were used as "observed controls." Because this assignment wasn't random, it suffered from many of the potential biases that perpetually worried Hill.

Fortunately, Francis's statistical lobbying at least partially won out. Though most states chose to enroll in the observed controls study, several opted into the randomized evaluation, including California, Illinois, Massachusetts, New York, Ohio, and Francis's home state of Michigan. A cohort of 400,000 children were vaccinated totally at random. In this group, there were 33 cases of paralytic polio in the treatment group and 115 in the control group.[19] There were 71 percent fewer cases of paralytic polio in the treatment group than in the control group.[20] Regardless of how you reported the numbers, the chance that such a discrepancy would be seen if the vaccine did nothing was effectively 0.

Statistics like these conveyed unambiguity, and the press reports declared the vaccine an unmitigated success. However, the true effectiveness was more subtle.[21] Now we get into the hard part of using single endpoints to understand the outcome of randomized trials. Effectiveness against paralytic polio was different from effectiveness against nonparalytic polio. The vaccine's effectiveness against different strains of polio was difficult to ascertain. There was a large variability in the potency of the vaccines in the trial, leaving the appropriate dosage for future vaccination unclear.

Most importantly, the trial did not rule out whether there were cases of polio in the treatment arm *caused* by the vaccine. The Salk vaccine worked by taking live polio virus and "inactivating" it using a chemical procedure. Based on his calculations, Salk was convinced that the potential of any live virus remaining in his vaccine was nil. But after the field trials, demand for the Salk vaccine was through the roof, and either due to manufacturing variability or carelessness, a batch of vaccine manufactured at Cutter Laboratories was released with large quantities of live virus, resulting in at least 250 cases of polio.[22] This is a quarter of the total observed polio cases in the field trials. After the incident, the Salk vaccine was withdrawn from the market, and public distrust of the vaccine and the scientific establishment festered.

The randomized trial of the Salk vaccine was simultaneously an undeniable success and a cautionary tale. It set a path forward for randomized testing of vaccines to ensure safety and efficacy. Yet, it showed that randomized trials could do only so much in understanding how a vaccine could impact future waves of illness. Randomized trials capture an instant in time. There is a population with a particular immunity profile to disease. There is a particular amount of illness in circulation. There is a particular method for manufacturing a vaccine. When these particularities change, the results of the RCT may no longer apply.

Contemporaneous with the development of new vaccines, breakthroughs in chemotherapy emerged in the late 1940s, promising the cure of many formerly terminal cancers. Given the promise of these treatments but also their evident toxicity, the National Cancer Institute became a natural home for innovations in clinical trials in the United States. They led the charge to bring Hill's trial designs more broadly to the United States. After significant lobbying, the NCI was flush with cash from Congress but daunted by the long list of cancers and potential therapies. Statisticians at the National Cancer Institute developed programs to run multicenter, multistage trials to establish a national knowledge base for fighting cancer.

Michael Shimkin, the head of NCI's Biometry and Epidemiology Branch from 1956 through 1963, was especially devoted to Hill's methods. While under his leadership, the NCI helped organize a variety of cancer trials, databases, and epidemiological surveys. One of his more audacious and impactful projects was a large-scale investigation into the effectiveness of mammographic screening for breast cancer.

NCI had been concerned with the effectiveness of drugs and chemicals, but Shimkin thought that biometrics could bring understanding even further. If clinical trials could evaluate preventive measures like vaccines, they should also apply to non-pharmaceutical preventive measures. Interventions that might potentially reduce incidence of or death from cancer. Among many possible preventives, Shimkin set his sights on *screening* for cancer. If doctors routinely examined their patients for cancer, could earlier diagnosis help reduce cancer deaths?

With new developments in radiology, Robert Egan developed a new method to examine breast tissue for cancer.[23] Egan had found a repeatable means of X-ray scanning women's breasts with relatively low radiation. In the 1950s, the standard practice for detecting breast cancer was monthly self-examination for lumps. Egan's mammogram, in theory, could find these lumps earlier and hence allow for an earlier intervention. This was a hypothesis that needed testing. As Shimkin put it, "The procedure would have limited value unless it could be shown that the cancers thus detected were indeed more curable."[24] Detection by a mammogram was to be followed by a biopsy, and the tumor would be surgically removed if deemed necessary.

The promise of cancer screenings like mammography is based on the principle of early detection. It seems plausible that if a patient can detect their cancer early, then their treatment can be less extreme. Many cancers begin as small localized tumors, growing in a specific part of the body. As they grow larger, operations to remove them become more invasive as the tumor impinges on more healthy tissue. And if the cancer metastasizes, it invades so much of the body that it may be too late to treat. This conventional wisdom motivates a commonsense idea. If you can intervene early and remove cancer while it is small, you may be able to stop the cancer in its tracks with limited patient suffering. It also seemed plausible that the earlier tumors were detected, the less likely they were to have spread to other parts of the body. If the cancerous region was localized, the surgical procedure could remove it and potentially end a patient's cancer. Proponents of screening also worried that by the time a cancer causes symptoms that propel a patient to consult their doctor, it would already be in an aggressive stage. Finding cancers before they are symptomatic again suggests they will be more treatable.

All of this reasoning seems reasonable, but by 1960, there was already plenty of evidence that early detection had limited value in the treatment of breast cancer. Self-examination had been promoted for nearly thirty years, and physicians were removing smaller and smaller masses over time. However, despite the widespread promotion of earlier detection through these self-exams and updated treatments, the rate of breast cancer deaths in the population had not changed over two decades.[25]

This is when the mammogram entered the picture. Egan had found that his mammographic technique could find malignancies that were not detected by a standard breast exam. A flurry of reports soon followed, confirming that Egan's method was both easy to implement and more sensitive than a physician's breast exam.[26] However, none of these results came from trials.

Shimkin, who had become devoted to the cold, hard "Numerical Method," wanted to put these claims about mammography under the scrutiny of a rigorous clinical trial. This would scientifically *prove* that early detection was beneficial. But Shimkin's back-of-the-envelope calculations suggested these trials would be a daunting task. He determined the study had to be large enough so that there would be between 10 and 100 bad events in the control group. Although current estimates suggest that 1 out of 8 women will be diagnosed with breast cancer at some point in their lives, the fraction of women who have undiagnosed breast cancer at any given time is far smaller (https://seer.cancer.gov/statfacts/html/breast.html). To have enough cases to determine whether mammography had any impact, Shimkin calculated that his study would require 60,000 participants and at least five years. This trial would cost millions of dollars.[27] It would also need a clinical partner to oversee the massive trial. Shimkin found a dedicated and eager partner in Philip Strax.

Philip Strax was the director of radiology at City Hospital in New York. His wife had succumbed to breast cancer at the age of 39, and according to several people who knew him, her death deeply wounded him. In many ways, her death radicalized him to find a definitive cure for breast cancer. Strax was a radiologist, and given his training, it is not surprising that he conceived a cure through radiology itself.[28]

With funding from NCI, Strax was the lead investigator on Shimkin's mammography trial. Strax's team sought participants from enrollees in the Health Insurance Plan of Greater New York. The trial would later be referred to by the insurance plan: the Health Insurance Plan Study, or the HIP Study. The study was a massive undertaking. Between 1963 and 1966, the study enrolled around 62,000 women between the ages of 40 and 64, assigning every other subject into the treatment and control groups.

The control group was provided the standard of care at the time. The women in the treatment group were offered mammograms. If they accepted, they were invited for a follow-up mammogram the following year. If they accepted the second mammogram, they were invited for a third and final mammogram the year after that. It's worth pausing for a moment to think about what the randomized treatment *is*. In the HIP study, women were assigned to the *offer* of a mammogram, not a mammogram itself. This seems silly at first blush: We care about the power of the mammogram, not the offer. But you can't force someone to have an X-ray. The best you can do is discuss the costs and benefits with the patient and hope she is persuaded. And hence, the effect of the offer is the best we can hope to measure.[29] In the HIP study, plenty of women declined the offer of screening. Sixty-seven percent of the treatment group had one mammogram, and only 40 percent had all three. Screening trials are looking at a far more indirect intervention than the drug and vaccine trials we have discussed thus far. We might pose the trial question as, *"What is the effect of offering a woman a mammogram on whether or not she dies from breast cancer within five years?"*

Let's find out the answer. The study tracked the health of the participants for over 15 years. The first data released compared outcomes five years after women had enrolled in the study.[30] The first question you might ask is whether mammography offers reduced deaths. The answer was no: about 3 percent of the participants in both groups passed away within five years of enrollment in the study. The question to ask next is whether mammography offers reduced deaths attributable to breast cancer. This was harder to answer because the trial had so few breast cancer deaths. Only 0.2 percent of the control group died from breast cancer. As is always the case in these screening trials, because you are evaluating a healthy population, only a small fraction will ever be diagnosed with breast cancer during the observation period. An even smaller fraction will succumb to the disease. Though breast cancer is a horrible disease, and so many of us have been close to someone who has had it, fatal breast cancer in women of the ages tracked in the HIP study is less frequent than we commonly believe it to be.

Nonetheless, the picture starts to look more favorable for mammography once we focus on the cancer death data. In the five-year follow-up period, there were 40 cancer deaths in the treatment group and 63 in the control. This seems like a substantial reduction. Indeed, using our thought experiment, the odds that we'd see at most 40 heads if we flipped $63 + 40 = 103$ coins is about 70 to 1. Now, on the one hand, this is worse odds than in the streptomycin trial, and we saw that it wasn't a great treatment for TB. On the other hand, it's certainly a signal that should be taken seriously.

Since breast cancer is rare, the *absolute* difference in the death rates of the treatment and control groups is small, equal to about 0.07 percent. That's a tiny percentage. But for rare diseases, it is often helpful to look at the *relative* difference. The *ratio* (rather than the difference) of the rates in treatment and control is about 1.6. That is, the rate of cancer deaths in the treatment group is about 1.6 times lower than that in the control group. A 1.6 reduction feels like a significant number. Still, for over 31,000 offered mammograms and 65,000 actual mammograms scanned from 20,000 women, only about 20 cancer deaths were prevented. Even taking the study numbers at face value, roughly 1,400 women needed to be offered an uncomfortable procedure in which they were exposed to X-rays to prevent a single breast cancer death.[31]

The results of the HIP trial were decidedly mixed. As I will discuss momentarily, there were plenty of early criticisms of the study's evidence of benefit. But Strax was convinced well before the study was even completed. He became a mammography proselytizer. He opened a mammography center in 1968, barely after the *enrollment* for the study had completed. He wrote an audaciously titled popular book, *Early Detection: Breast Cancer Is Curable.*[32] I don't know what led Strax to make such a hyperbolic false promise, but it is clear there was intense excitement about the promise of screening. The inconclusive results of the HIP trial would not dull this excitement. Through a strong activist push in the United States, we transitioned from about 15 percent of women having mammograms in 1986 to in excess of 70 percent of women over the age of 40 having mammograms by 1999. Similar rates continue through the present.

But we have to come back to the original question. Strax told us that breast cancer was curable, and the question is, does cancer screening actually cure cancer? The evidence, from the beginning, was decidedly inconclusive.

———

Though Strax and his co-investigators in the HIP study were sure they had found a breakthrough, their trial was met with plenty of skepticism. The small but significant differences between the groups had many potential explanations. For instance, John Bailar noted that only 15 percent of the breast cancers in the treatment group were even discovered by mammography alone.[33] The others were found by either manual breast exams or manual exams in tandem with the mammogram. As noted in later analysis, the mammography techniques used by the HIP study were only sensitive enough to detect cancers about 40 percent of the time.[34] Bailar concluded that mammography could have only prevented at most 12 deaths in the HIP study, putting the number needed to screen to prevent one breast cancer death closer to 2,000. Moreover, it was clear that almost all the benefits of screening were for women over 50.

There was a clash between what the numbers said and what cancer advocates wanted to believe. The HIP trial showed no benefit for women under 50, but the American Cancer Society lobbied to encourage mammograms for all women 35 and older. The ACS rationalized that earlier screening couldn't hurt, and advocating for more widespread screens would create greater awareness about breast cancer in the general public. Shimkin, the trial architect, was dismayed, noting, "The set of considerations is different for the analytic research mind and for the proponent and activist of health programs. The proper point of equilibrium remains in the realm of social decisions."[35]

Mammography advocacy continued well after the HIP Study, with a significant push for broader adoption in the United States through the 1980s. Other countries also became interested, and clinical trials were launched in Canada, Scotland, and Sweden to see if the HIP results could be replicated—or even improved—with more reliable, modern screening technology.

Unfortunately, *more* trials didn't make the situation *less* confusing. The Swedish trials seemed to show considerable benefits for screening, but the Scottish and Canadian trials were inconclusive at best.

One of the popular ideas for sifting through the confusion was a *systematic review*. Systematic reviews carefully assemble all clinical trial evidence on a particular intervention and synthesize a general conclusion about that intervention's value. *The Lancet* published such a systematic review on mammography led by Peter Gøtzsche and Ole Olsen in 2000.[36] Gøtzsche and Olsen collected eight studies and combed through the trial data, looking for potential biases or errors that might complicate general findings.

Regarding HIP itself, Gøtzsche and Olsen flagged issues with how patients were randomized and excluded after being enrolled in the trial. The HIP project was never entirely clear about how patients were randomized into treatment and control, and the various trial reports gave different numbers about how many participants were in each group: 31,000, 30,000, 30,131, 31,092, 30,239 in the study group, and 30,756, 30,765, and 30,565 in control. The variance in these numbers was worrisome because the number of cancer deaths was so low. If you missed a few here or there, the effectiveness of the trial quickly diminished. Exclusion was the bigger worry: if the study was dropping high-risk women from the treatment group, this could bias the results in favor of screening. Reported numbers by the study suggested there could be more women in the control group who had previously had lumps in their breasts.

This led Gøtzsche and Olsen to further investigate the possibility that patients were improperly dropped from the study group. Patients were excluded from the HIP trial if they had a preexisting diagnosis of breast cancer, but it was not specified how to surmise whether they had breast cancer before the trial. According to the study staff, women in the control group "were identified through other sources as having had breast cancer diagnosed before their entry dates." The study documentation is unfortunately not specific about what "other sources" means. In complex trials like HIP, important probative details often go unwritten. Hence, the study authors conceded that prior breast cancer "was most completely ascertained for screened women."[37] In other words, women in the group given standard of care could have been entered into the study with a breast cancer diagnosis and never removed from the study, just because the physician never asked. This could have big consequences. If too many women with prior breast cancer were kept in the control group, it could wash out the tiny differences in breast cancer mortality altogether. Indeed, Gøtzsche and Olsen scoured the study data and found that 853 women were excluded from the treatment group but only 336 from the control group. They note, "If only 10% of these excluded breast cancer cases are added as breast cancer deaths per 18 years of follow-up, the breast cancer mortality becomes higher in the screened group than in the control group, since the difference in breast cancer mortality at that time was 44 deaths."[38] Defenders of cancer screening put forward various plausible arguments to explain why this ascertainment problem wasn't an issue, but this required leaning on assumptions that were not directly evaluated in the trial.[39]

The question of uneven ascertainment led to an even trickier question: What does it mean for a death to be *caused* by breast cancer? What is a "breast cancer death"?[40] Usually, in a trial, experts are required to examine medical reports and deliberately rule out other causes. In the HIP trial, 72 percent of the death attributions were unblinded. Again, a staff with the bias of wanting to show screening worked might, perhaps, subconsciously be more willing to attribute a death to breast cancer in the control group than in the treatment group. That is why, in modern cancer trials, such assessments are made by blinded third-party clinicians with no interest in the studies.

Now, many of these discrepancies could be explained away. Perhaps they were only inconsequential bookkeeping issues. But small bookkeeping issues can sink large trials! One hundred bookkeeping errors in a trial with 100,000 people might seem small. But it is substantial and damning when that trial is looking for a difference of 10 cases between the treatment and control groups. Rare events demand impossibly pristine trials. When comparing the differences in the screened and unscreened groups, small details about bias in endpoints, errors in randomization, exclusion and ascertainment issues, and causal attribution of mortality can quickly compound and suggest effects that are not present.

HIP was an early trial and had many practices that are no longer accepted.⁴¹ Did the later trials bring more clarity? Unfortunately, the seven other mammography trials had problems of their own. Gøtzsche and Olsen found issues with the randomization schemes, reporting of cohort size, and cancer ascertainment in 5 of the 7.

That left Gøtzsche and Olsen with only two trials clearing the bar for bias, one from Sweden and one from Canada. These two trials showed no benefit for mammography at all. According to Gøtzsche and Olsen's analysis, once a trial was properly randomized and biases from ascertainment and cause of death assessment were removed, there was no health benefit associated with screening for breast cancer.

The twenty-five years since Gøtzsche and Olsen's analysis have not brought clarity to the value of breast cancer screening. On the one hand, in 2014, the Swiss Medical Board recommended sunsetting mammography screening programs.⁴² On the other hand, in 2023, the United States Preventive Services Task Force lowered their recommended age for mammography to 40. This US recommendation wasn't based on new data or trials, but rather on a speculative computer model and a call for health equity.⁴³

Given the numerous randomized trials, why is there such confusion and disagreement? It's worth looking at the arguments from Nikola Biller-Andorno and Peter Jüni, two members of the Swiss Medical Board who wrote the controversial report on mammography. First, it's hard to continually do trials on the same interventions over and over again, but the world is very different today than it was in 1963. Cancer treatment has significantly improved since the dawn of mammography. Many *symptomatic* breast cancers are now treatable with a combination of surgery, radiation therapy, and pharmaceuticals. Because of improved treatment, the benefits of early detection are muted.

Second, medical consensus now accepts that different types of cancer look similar on an X-ray. Some cancers grow aggressively, and some grow slowly. Both look the same in a mammogram. For the most aggressive cancers, even mammographic detection is too late, and the cancer may quickly return and metastasize after surgery. Some cancers grow so slowly that they will never be a health problem, and the harms of surgery and treatment outweigh the benefits. Screening can't distinguish between these cases, and this limits its utility.

And finally, the HIP study design itself may have created an illusory perception of mammography's effectiveness. There is substantial evidence that the population at large has been convinced that mammograms themselves are effective at increasing overall cancer survival. A survey of women in the United States, the UK, Italy, and Switzerland found that women's perception of the benefits of mammography dramatically outweighs the actual benefits.[44] Seventy-two percent of the American women surveyed believed that mammography reduced the risk of breast cancer deaths by half or more, and that 80 or more deaths would be prevented per 1,000 women invited for screening. In reality, even with the most generous readings of the accumulated RCT evidence, mammography improved survival by 20 percent at best, and only approximately one death per 1,000 women was prevented. This gross misperception of risk and benefit led to more mammograms, more false positives, and more unnecessary procedures with little benefit to patients' well-being and overall survival.

The popularity of early cancer detection has waned as the same issues of overdiagnosis and diminishing effectiveness arose in other cancers as well. Widespread prostate cancer screening was a notable failure, as the false positive rates were so high and the undue harms of overdiagnosis, including increased rates of suicide, so pronounced.[45] Despite the diminished zeal for cancer screening, the rate of cancer deaths continues to drop.[46] This sort of anticorrelation between screens and cancer shouldn't be ignored, even if it wasn't studied in a proper medical trial.

The one cancer screening that remains widely popular is the colonoscopy. Colonoscopies not only look for precancerous lesions in the colon but also remove them. Since we think colon cancer must arise from these precancerous regions, removing them early must improve outcomes. In fact, the US Preventive Services Task Force recommended in 2020 to lower the age of colonoscopy in all adults from 50 to 45.[47]

But what is the evidence that colonoscopies prevent colon cancer deaths? Believe it or not, no RCT was done on colonoscopy until 2009.[48] Let me read off the results of the study in the cold numerical fashion that Michael Shimkin would have desired: 85,000 participants from four countries aged 55–64 were randomized to the invitation to a colonoscopy; 42% of those offered accepted the screening, and the patients were observed for ten years. About 0.28% of the patients offered a screen died of colon cancer. About 0.31% of the control group died of colon cancer. Eleven percent died from something else. The difference between the groups was not deemed statistically significant, but you may draw whatever conclusions you'd like from these statistics.

I doubt Bradford Hill anticipated the widespread embrace of the randomized clinical trial. Today, randomized trials are considered the pinnacle of causal evidence in medicine. Randomized trials are the backbone of the FDA's three-phase approval process for new pharmaceuticals. Trials are also required for many medical devices. There are tens of thousands of randomized clinical trial reports archived by the Cochrane Library, available to help find indications of countless treatments.[49]

Randomized trials have also proven a powerful tool for finding what *doesn't* work in medical practice. When examined by the sharp lens of the randomized trial, conventional wisdom often buckles. A notable example is the CAST randomized trial, which showed that certain drugs designed to dampen unusual rhythms in the heartbeats of patients post–heart attack led to increased mortality.[50] Common sense had told cardiologists that these irregular heartbeats were harmful, but a trial showed this common sense was completely wrong. Similarly, the Women's Health Initiative Trial showed that while estrogen therapy could be beneficial to treat early symptoms of menopause, it increased the risk of cardiovascular disease in older women.[51] The non-randomized data collected before the trial had actually pointed in the opposite direction, suggesting that hormone therapy was preventive of long-term disease.[52] The systematic, large, randomized trial pointed to unseen risks and problematized a broad recommendation for all women. Women and their doctors had to discuss the potential risks and benefits before starting estrogen.

With their successes in medicine, randomized trials have grown popular in diverse fields beyond healthcare. Randomized trials have revolutionized experimental social science, helping to bring a "credibility" revolution to economics starting in the 1990s.[53] Economists found the logical rationality of the randomized trial to be more compelling than their prior econometric data analysis, which was often disregarded as "correlation not causation." Development economists now run giant field trials testing whether mosquito nets reduce childhood mortality,[54] whether giving children deworming pills reduces poverty,[55] or whether providing poor children glasses improves their performance in school.[56]

Randomized trials have had the most buy-in in the tech industry. The *A/B test* has emerged as a go-to scheme to evaluate whether human-facing internet technology works. A/B tests are just randomized trials, but with software features. Tech companies can run randomized trials showing part of their user base a new feature and controlling with another randomly selected set of people. Based on this comparison, they can track various metrics to determine whether or not it's worth shipping this new piece of code.

The examples in the social scientific settings here highlight the problematic epistemological value of randomized experiments. Think about the deworming pill example. An influential randomized study in Kenya found that giving all children a deworming pill, whether they were ill or not, improved health and scholastic aptitude.[57] Yet trying to establish a causal link between a deworming medication and school performance is even more tenuous than linking an offered screening to a cancer death. As the intervention gets more and more decoupled from the outcome, too many additional explanations can come in to wash away conclusions, and it's a bit too easy to try to control for these explanations in analysis and end up introducing data processing errors. Such errors plagued the deworming trials, with several papers calling the main results into question.[58] The results of the initial big study have failed to replicate elsewhere.

Moreover, the eyeglasses example should also give us all pause. Why are we trying to determine whether giving myopic children glasses improves their outcomes? This is a grotesque question. I can think back to my own time in middle school when my chronic headaches were cured by a simple glasses prescription. Of course we should give poor children glasses if we can. In their dismal way, economists were interested in knowing if the cost of two dollars per child was worth the downstream economic benefits, whatever they may be.

What about in technology? Some engineer comes up with some fancy new widget that they think will increase engagement on their app's social timeline. Because app companies control every aspect of what you see in their app, they can ship different versions of their software to different people. They can randomly assign 5,000 people to see the new widget and 5,000 people to serve as controls. Watching the interactions of these people with the app for a few weeks, they can measure some notion of engagement and see if it goes up. They can also track other critical metrics that they need to not go down and make sure the widget doesn't harm those metrics. If successful, they can conclude that the new widget improves their service and ship it to the entire user base.

The A/B test is just a randomized clinical trial with far lower stakes. Software companies opt you into these tests whenever you click "agree" to use their apps. Once they have you, they'll run all sorts of tests. They can test if changing the shade of blue they use on the app leads to increased revenue. You think I jest, but Google once bragged about having A/B tested over 50 shades of blue.[59] Perhaps more practically, A/B tests will examine whether some advertising campaign increases clicks. They can test if adding more ads to your feed increases the amount you click on ads. They can test whether redesigning the organization of the page gets you to spend more time on the website. They can test which dance routines get people more addicted to scrolling.

To use any tech product, we all have to agree to some terms of service. And part of that agreement is to let tech companies treat us all as lab rats in any of their desired experiments. A company can run thousands of A/B tests a week if they're ambitious. The cost of the test is essentially zero. If the A/B test were a powerful tool for moving policy, we'd see it in industry. The story is, of course, far more complicated.

The widespread application of randomized trials has exposed their inherent limitations. In particular, there is always a question of whether the effects hold up after the trial. When we do a randomized trial, we get a sense of the impact *within* the study population. Because assignments are randomized, we remove many potential sources of biases in determining the effectiveness of a treatment. But what does a randomized clinical trial tell us about the next patient who walks in the door? If twice as many people in the treatment group recovered than in the control group, does that tell us that this person is "twice as likely to recover" if they receive the treatment? What does that mean? We make an internal inference about how this patient is similar to those in the study. Is this valid? And for how long is this valid?

As we saw in the case of cancer screening, multiple studies with slightly different settings, protocols, and evaluations can suggest a broad spectrum of potential benefits. Actual benefits may strongly depend on context. And benefits might only be good for so long. I described how improvements in cancer treatments would dramatically reduce the benefit of a cancer screening.

Understanding how results of a study transfer to new contexts is challenging, and it's a challenge not at all restricted to the medical domain. The track record for RCTs in other areas isn't much better. The famously data-driven Google quickly became aware of the perils of leaning too heavily on A/B testing. Google users would adjust to the changes made by the company, rendering the effects marginal.[60] There is nothing that can prevent a series of experiments from simply running in circles. A configuration deemed less profitable a month ago might now be more profitable than what is currently shipping to customers. Data scientists have found that changes must be consistently evaluated, demanding even more resources to ensure current practice remains best practice. Experiments and data, it turned out, were not substitutes for thoughtful design.

A/B tests are also very sensitive to how the experimental outcome is defined. In a cancer trial, specifying the trial outcomes seems reasonably straightforward: You want to prevent deaths from cancer. Formulating a clean outcome is much more challenging in an A/B test. Think about trying to test features to boost "engagement." Engagement can mean all sorts of things, and measuring a client's happiness with a product is messy. If I'm trying to recommend articles for people to read, I might expect read time to be a good way to measure engagement. However, intervening directly to increase read time could ruin a person's experience. It might take them longer to read because the page format is less readable. That isn't good. Understanding what to measure is critical.

Even apparently simple outcomes like revenue are hard to evaluate with A/B tests. In computational advertising, Randall Lewis and Justin Rao showed that most experiments that tried to estimate the value of advertising campaigns couldn't tell the difference between a profit and a loss. They calculated that properly evaluating impact would require A/B tests with millions of individuals. Such scale was impossible even for the largest tech companies.[61]

Similarly, impact evaluations in development economics seldom generalize to new populations. Economist Eva Vivalt looked at over 600 studies in development economics and found that the estimated impacts varied wildly. After a careful reevaluation, many interventions initially measured as helpful turned out to be harmful. The actual deployments of programs by governments tended to have considerably smaller impact than those measured in the study. Among over thirty interventions, Vivalt found only two were consistently beneficial. The first was providing bed nets to prevent malaria. The second was an intervention to reduce truancy: giving poor families money if their children attended school.[62]

If these results are so fragile, then is there a value in randomized experimentation at all? Looking back to their earliest motivations, the answer has been staring at us all along. Randomized trials are a means of regulation.

The preposterous eyeglasses trial reinforces that the main goals of randomized trials are policy evaluation, not scientific identification of cause and effect. The question is rarely "Does this intervention work?" but rather "Is this intervention worth it?" From its inception, the FDA demanded a regulatory mechanism to test whether drugs were safe and effective before making them legal. They wanted to know if the benefits outweighed the risks. In development economics and social science, the goal of a trial is to determine whether a government should allocate funding toward this intervention. Would the population benefit enough to justify the cost? And, fascinatingly, in technology companies, randomized trials are now widely used to determine whether it is worth adopting pieces of code.

Without A/B tests, it's tough to know if a feature does what you want. Any other evaluation framework that people have tried has too much measurement bias. Software engineers only get promotions when they ship code, so everyone is incentivized to generate as many new features as possible. Some guardrails are needed, and better diagnostic tools can only help. A/B tests prevent "really wrong" things from screwing up code bases. This sounds an awful lot like how the FDA regulates drugs.

And maybe this shouldn't be too surprising. Researchers only run a study if they expect their treatment has an effect. They'd have seen it in earlier experiments and case series. They have a plausible mechanism in mind before they begin. The RCT is just the final hurdle used to convince regulators to approve their intervention. These regulators might be the FDA. They might be some government considering a new policy. They might be managers at a tech company.

This pragmatic view of randomized controlled trials is how they are most often used in medicine, social science, and technology. Our modern FDA has a concrete set of rules about what burden of proof a drug company must demonstrate for a new drug. We can argue whether they are too lenient or too harsh, but they are clear. A tech company sets some threshold of "lift" for a new feature to demonstrate in an A/B test before shipping the new code. A country may choose to employ a new policy only after a field trial demonstrates the accuracy of a cost-benefit analysis. All of these trialists are measuring the effect of their interventions to make decisions. These decisions will be informed by what they measure in their randomized experiment. They'll consider the costs and potential harms of the intervention and decide if the measured benefits are enough to justify the costs and harms. As we saw in this chapter, this does not rule out the findings being ambiguous or wrong. The randomized trial provides a set of rules for making hard decisions, but it doesn't promise that the decisions will always be correct. No one can promise that. Randomized trials might not be perfect, but they give a set of rules for informing these policies, and sometimes a clear set of rules is better than nothing.

5

When Past Performance Is Indicative of Future Results

IN HIS landmark 1948 paper "A Mathematical Theory of Communication,"[1] Claude Shannon made perhaps his most significant mark on the information age, laying the foundations for digital communication. Shannon described how to mathematically model the transmission of information over unreliable transmission links. With the proper encoding of their message, people could communicate perfectly over long distances, even in the presence of all sorts of noise, corruption, and interference. We have Shannon to thank for the ability to stream TikTok from anywhere on Earth. Shannon's paper was unfathomably deep. As was characteristic of his work, this single paper launched multiple research fields: communications theory, information theory, and, notably, *natural language processing*.

Why language processing? Shannon observed that if a signal has some redundancy, we can encode the information into a smaller representation, send it through the channel, and decode it on the other side. Encoding and decoding would mean fewer bits sent through the channel and, hence, faster communication times. But the key to encoding was the redundancy. If signals were *predictable*, we could transmit them more efficiently.

Shannon wanted to know how predictable English text was, and this required devising an encoding for text. Each character in the alphabet plus a character for the space between words can be encoded with five binary bits. This is because there are 32 different strings of ones and zeros with a length of five, but only 27 characters. We could let A be denoted by 00000, B by 00001, C by 00010, and so on. Shannon wanted to find the smallest possible encoding.

To determine the redundancy of text, Shannon built models to simulate it. His models used probability: He would choose the next character or sequence of characters by sampling from some probability distribution. What was a good distribution? Shannon decided to use *Markov processes*. In Markov processes, each symbol in a sequence is generated from a probability distribution. Each symbol's probability only depends on a previous window of text. The last set of characters thus predicts the next character.

Shannon presented the simulation of several different approximations of English. The most basic approximation of English generates characters completely at random, each with probability 1/27:

XFOML RXKHRJFFJUJ ZLPWCFWKCYJ FFJEYVKC-QSGHYD QPAAMKBZAACIBZLHJQD.

This doesn't look like language at all. There are too many consonants. We know letters don't all occur with the same frequencies. If we built a model that generated letters at random but with probability equal to their frequency in written English, we get text like this:

OCRO HLI RGWR NMIELWIS EU LL NBNESEBYA TH EEI ALHENHTTPA OOBTTVA NAH BRL.

Better, but still not English. This language model did happen to predict the slang "NAH" as part of the text. But strings "LL" or "BRL" are certainly not words nor approximations of words. To get more believable character sequences, Shannon built a model where the next character's probability equaled how frequently it occurred after the previous three letters. With this, he got text that looks like Latin:

IN NO IST LAT WHEY CRATICT FROURE BIRS GROCID PONDENOME OF DEMONSTURES OF THE REPTAGIN IS REGOACTIONA OF CRE.

Shannon got progressively more sophisticated in his modeling, switching from characters to words. Sampling words according to their frequency in common corpora yielded this passage:

REPRESENTING AND SPEEDILY IS AN GOOD APT OR COME CAN DIFFERENT NATURAL HERE HE THE A IN CAME THE TO OF TO EXPERT GRAY COME TO FURNISHES THE LINE MESSAGE HAD BE THESE.

But when he took things a step further and treated the words as depending on each other, something magical emerged. Here's his text where the probability of the next word was equal to the frequency with which it occurred after the previous two words. The passage starts to look like English:

THE HEAD AND IN FRONTAL ATTACK ON AN ENGLISH WRITER THAT THE CHARACTER OF THIS POINT IS THEREFORE ANOTHER METHOD FOR

THE LETTERS THAT THE TIME OF WHO EVER TOLD THE PROBLEM FOR AN UNEXPECTED.

After designing language models, Shannon proceeded to show that the right way to measure redundancy in text was in terms of prediction errors. Highly redundant text, like a long chain of the same character, is easy to predict. Complex and uncompressible text (Shannon used the example of James Joyce) was unpredictable. If prediction measured redundancy, then the best way to measure the information content in text was to build a really good model of it and then estimate the average number of times that model made incorrect predictions. Shannon would measure his prediction errors using a metric called *conditional entropy*.[2] Entropy, a measure of disorder from statistical mechanics, quantifies the number of bits required to specify the next character from the previous characters. Conditional entropy determines how compactly messages could be compressed when transmitted.

But how much information was in language? Shannon tried to estimate this in his second paper, "Prediction and Entropy of Printed English."[3] He computed the information content in his simple language models. For the naive language model with spaces, it was 4.8 bits per letter. Using letter frequencies, this number went down to 4.0 bits per letter. If two letters were used to predict the next, the number was closer to 3.3. Using about 100 characters, Shannon estimated that the entropy was around 1 bit, somewhere between 0.6 and 1.3. Seventy years later, using thousands of tokens to predict the next, this estimate seems about right. Recent work shows this number is around 0.7 bits per character.[4]

Shannon proposed many things before 1950:

1. Language is predictable. Written English is approximately 85% redundant.
2. Language can be modeled by predicting the next character, word, or token based on the prior text.
3. The way to build these language models is to analyze many corpora of English to fill in the right statistics.
4. The way to evaluate the quality of approximation is through conditional entropy.

These four bullet points underlie the engineering of today's impressive large language models. To get our modern AI chatbots, we need only combine Shannon's language models with an algorithm called *stochastic gradient descent*, some clever ways to compactly represent probabilities of sequences with artificial neural networks, and the terabytes of text out there on the internet.[5] The stochastic gradient method, an optimization method like gradient descent that could find solutions in spite of noisy information, would be invented by Herbert Robins and Sutton Monro in 1951.[6] Artificial neural networks were proposed by Warren McCulloch and Walter Pitts in 1943[7] and studied throughout the remainder of the 1940s and 1950s. So why did it take us so long to get to the modern AI chat systems (like OpenAI's ChatGPT) that became the buzz of the 2020s?

One obvious answer that we have seen in the previous chapters is that we needed to wait for computers to speed up. Shannon did his work on language modeling at Bell Labs during World War II before computers existed. Much of the work I'll describe in this chapter was done on computers less powerful than the controller in a microwave. Bringing computation to these seminal techniques has led to mind-blowing technology.

But there was also something more subtle that kept us from realizing today's AI chat systems earlier. The technology of prediction is frustratingly hard to formalize. I won't write up a graveyard of ideas that have been (or should be) abandoned in the applied predictive sciences of pattern recognition and machine learning, but know I could pen a whole other book about them! In this chapter, I'm going to focus on what worked. I'll take an idiosyncratic path through the history of data-driven prediction, highlighting how the most essential parts of the field of machine learning were not necessarily the most technical or mathematically profound. What fuels progress in engineering prediction is faith in predictability itself. A faith that the future can be predicted from a sufficient number of observations of the past.

Shannon's methodology for building language models is an example of what we now call *machine learning*. More specifically, it is an instance of *statistical pattern recognition*. The same statistical pattern recognition powers most modern artificial intelligence software. Let me take a moment to describe how it works.

To frame the prototypical machine learning problem, I like to think about a hypothetical spreadsheet. Each row of the spreadsheet corresponds to some unit or example. But I don't care what the units mean. I just know that I have a bunch of columns filled in with data. And I'm told one of the columns is special. I am about to get a load of new rows in the spreadsheet, but someone downstairs forgot to fill in the special column. Management has tasked me with writing a formula to fill in what should be there. For whatever reason, I don't get to see these new rows and have to build the formula from the spreadsheet I have. The formula can use all sorts of spreadsheet operations: It can assign weights to different columns and add up the scores, it can use logical formulas based on whether certain columns exceed particular values, it can divide and multiply. Microsoft Excel has nearly 500 different functions for you to combine as you see fit.

How might I go about finding the right formula? I can try to use the values I've seen so far in the special column. I could imagine matching rows. If I take a new row, compare it to all of the others I have so far, and find an exact match, I can use the value of the special column in the matched row as my value in the new row. That's pretty straightforward. I'll look for duplicate rows where I have data and fill in the missing values by using the values in those rows.

But what if I find multiple duplicates with different numbers in the special column? What do I fill in now? I could use the most frequently occurring value. I could use the average value. I could return a function that spits out a random value every time I open the spreadsheet. All of these are valid options, and I'll have to ask management what would be best. For this, they'll hopefully give me a precise specification that lets me define a quantitative measure of accuracy, and I'll pick the value that maximizes accuracy.

And what if no row matches exactly? Now I have to make another decision. I could find the closest row and predict the value listed there in the special column, but closest in what sense? Maybe I should look at dissimilar rows and use the opposite value of what they have. There are so many options for what I can do, and choosing the formula now seems impossible.

I'll do an experiment. I'll take the last row of my spreadsheet and pretend I don't have the special column. I'll write as many formulas as I can. The formula might need a very complicated expression, but you can do arbitrarily absurd things in

Microsoft Excel.[8] Out of all these formulas, I'll pick one that guesses the special column. Then, I'll use that formula for the new data, too.

But why single out that last row? I can do something similar for every row! I'll invent a set of plausible functions. I'll evaluate how well they predict on the spreadsheet I have. I'll choose the function that maximizes the accuracy.

This is more or less the art of machine learning.

Though it sounds preposterous, I promise you that it is no different from what modern data scientists do. They are given databases and tasked with maximizing accuracy in predicting special columns. Let me describe some famous examples I use when teaching undergraduates. The columns could be lengths of parts of a flower, and the hidden column could be the flower species. The columns could be measured attributes of a tumor, and the hidden column could indicate whether the mass is malignant or benign. The columns could be a bunch of attributes of neighborhoods in Boston, and the missing column could be the median home price.[9] Thousands of budding machine learning engineers have seen that the missing columns are predictable using the procedures I described. They seldom even need to think about what the entries mean.

Even the random number generator joke above is a legitimate description of machine learning practice: Shannon's language prediction game would correspond to the columns being characters, and the goal, returning a sample from a language model. And since they are just absurdly large, computationally intensive versions of Shannon's models, the metaphor extends to contemporary Large Language Models as well.

Given that prediction software is so powerful in building compelling language models, you can probably envision even more powerful applications of this technique. What if you had spreadsheets of lab tests of all sorts of people? Could you predict who might develop cancer? What if you had a massive dataset of amino acid sequences and protein structure information? Could you predict protein structure from the sequence alone? What if you had a spreadsheet of instrument data from cars driving all over the United States? Could you predict how a car should act and build a self-driving car? The admission fee for machine learning is believing that this kind of pattern recognition is possible. You must accept that the special column is predictable from the rest.

The question then is only about computation. If we assume a pattern is well predictable by some computational mechanism, can we find that computation? When does searching for a function that maximizes accuracy on a given dataset yield a function that makes good predictions on data we haven't seen yet?

We still don't have satisfying answers to these questions. But we have seventy years of machine learning practice that we can study and mine for its own patterns. Data gathering, function representation, and search have been the backbone of machine learning since the late 1950s. Let me first describe why engineers think computational pattern recognition should be possible. I will then give the first example of someone successfully putting the prediction paradigm to use.

To understand why engineers in the early computer age believed in the potential of pattern recognition, let's first understand why Shannon thought it possible.

> How is an information source to be described mathematically, and how much information in bits per second is

produced in a given source? The main point at issue is the effect of statistical knowledge about the source in reducing the required capacity of the channel by the use of proper encoding of the information. In telegraphy, for example, the messages to be transmitted consist of sequences of letters. These sequences, however, are not completely random. In general, they form sentences and have the statistical structure of, say, English. The letter E occurs more frequently than Q, the sequence TH more frequently than XP, etc. The existence of this structure allows one to make a saving in time (or channel capacity) by properly encoding the message sequences into signal sequences. This is already done to a limited extent in telegraphy by using the shortest channel symbol, a dot, for the most common English letter E; while the infrequent letters, Q, X, Z are represented by longer sequences of dots and dashes. This idea is carried still further in certain commercial codes where common words and phrases are represented by four- or five-letter code groups with a considerable saving in average time. The standardized greeting and anniversary telegrams now in use extend this to the point of encoding a sentence or two into a relatively short sequence of numbers.

We can think of a discrete source as generating the message, symbol by symbol. It will choose successive symbols according to certain probabilities depending, in general, on preceding choices as well as the particular symbols in question. A physical system, or a mathematical model of a system which produces such a sequence of symbols governed by a set of probabilities, is known as a stochastic process. We may consider a discrete source, therefore, to be represented by a stochastic process. Conversely, any stochastic process which produces a discrete sequence of symbols chosen from a finite set may be considered a discrete source.[10]

Shannon's model assumes words in sentences are random. What does this even mean? The key is given in his second paragraph. He finds that language is well approximated by a *stochastic process*, a technical mathematical term for a sequence of random numbers. Why was Shannon thinking about modeling word sequences as random?

When we speak, we have the intention to communicate something. If anything, intentional is the opposite of random. But the people we speak to don't know what we're trying to say. They are not mind readers. For every start of every sentence, there are many possible ways it can end. Noises in the background might make it hard to hear some of the speaker's words. A good listener should be able to process many potential sentences under different conditions.

The demands on the listener give us three different paths to randomness in communication. First, the fact that sentences can be completed in multiple ways implies a non-uniqueness that we can associate with randomness. Each potential sentence completion has a different likelihood in the listener's mind. Second, noise can be modeled as random, and perhaps it's useful as a model to understand how much noise a communication system can tolerate. Third, since a system has to recognize many possible sentences, we can evaluate listening comprehension based on the system's average behavior.

In the 1940s, all three of these views of randomness were used to design new state-of-the-art communication systems. Shannon realized that if language was predictable, then words that were hard to understand because of background noises or unusual cadences and pronunciations could potentially be filled in by considering their probabilities in a language model. Moreover, if the language could be encoded before sending, codes could be selected so that less information needed to be sent when the next word was determined by context.

Removing noise had been a persistent issue in electronic communication since its inception.[11] By the 1920s, engineers had realized that electronic noise in communication systems was often well approximated using equations from statistical mechanics.[12] This was a powerful realization. Though the particular sequences of noise voltage levels were unpredictable, the average intensity of the noise was easy to characterize. With such characterizations, engineers could attempt to get around the noise either by increasing the intensity of their signals or by cleverly encoding their signals.

Norbert Wiener was one of the pioneers of this view of signal coding and prediction. Wiener additionally declared our third motivation for the random model of communication signals. Wiener acknowledged that communication methods had to work on a vast array of potential signals. And he decided that a communication system should be evaluated by how it performs on average over all such signals.

> No apparatus for conveying information is useful unless it is designed to operate, not on a particular message, but on a set of messages, and its effectiveness is to be judged by the way in which it performs on the average on messages of this set. "On the average" means that we have a way of estimating which messages are frequent and which rare.[13]

For Wiener, the ensemble gives us a metric against which we can optimize. We know in advance what we'll be tested on and evaluate our performance on average.

Wiener's evaluation schema enabled something unexpected. With the technology of the time, getting every signal perfectly correct wasn't achievable. However, accepting the possibility of error allowed engineers to design systems that could be compared. Each engineer could compute their average case performance, and other engineers could find clever innovations to improve that performance. Iteratively working with improved technology, there was hope of eventually getting the communication errors to zero.

In all three cases I list here, randomness was introduced into communication theory as a mathematical tool to mitigate different manifestations of non-uniqueness. Rudolf Kalman, a pioneer in automatic control whom we encountered in the optimization chapter, was never happy with this conflation of randomness and non-uniqueness. Kalman would later claim non-uniqueness is the central feature of all random processes.[14] According to Kalman, an outcome is random if, up to a set of equivalences on the outcomes, the future cannot be specified to a unique outcome. I tend to agree with his definition. That *some* horse wins a horse race is not random, but *which* horse wins the race is. That a die will stop rolling is not random, but which face it lands on is random. In a similar vein, the next word in most sentences is also non-unique. As a matter of mathematical convenience, modeling sequences as random lets you do certain calculations cleanly and report errors over averages of potential occurrences.

After Shannon and Wiener, data was modeled as naturally random whether it was justified or not. Randomness was too useful to question. The stochastic modeling paradigm was empowering across multiple fronts. Randomness let Shannon determine the maximum possible communication rate of diverse communication channels. Randomness let people design optimal filtering devices to remove transmission noise. And randomness let engineers evaluate systems based on how frequently they got answers correct.

If data in the future is random, that means data in the past was random until we saw it. We could perhaps use the idea that we were surprised in the past to mitigate how surprised we'll be in the future. This opened the door for computational pattern recognition.

Let's dig into why treating the past as random lets us predict the future. We're all familiar with the conventional wisdom of how to transmute counts of past events into likelihoods of future events.

- 16 out of 100 die rolls landed on a 6. Therefore, my chance of rolling a 6 with this die is 16%.
- Steph Curry has made 257 of his 281 free throws this season. Therefore, the probability he'll make a free throw in tomorrow's game is 92%.
- 70 out of the 127 2-point attempts in football last year were successful. Therefore, the probability of converting the next 2-point conversion is 55%.
- It rained in the morning three out of the last five days. Therefore, the probability it will rain this morning is 60%.
- In clinical trials, 900 of the 1300 people who took Ozempic lost over 10% of their body weight in 5 years. Therefore, I have a 70% chance of losing over 10% of my body weight by taking Ozempic.

We do this sort of count-to-chance conversion all the time. We collect a bunch of events, count the number of times they occur in the past, and then turn this frequency into a probability in the future. What do we need to be true for this conversion to be correct?

Intuitively, we'd like to assume that, all else being equal, each of these isolated events is identical in some capacity. We conceive each event to be a realization of the same process with some additional variability that we hope will average out if we collect enough data. Moreover, we think of these events as isolated from each other so that the order in which the events occur doesn't matter. All that matters is frequencies. Finally, we have to assume that each event is "random" and hence mentally equivalent to some game in a casino.

We use a few standard models in statistics to capture these intuitions. The first is that the events are *independent and identically distributed*. Statisticians use this model so much that they abbreviate it without periods as "iid" (though it's still pronounced "eye-eye-dee," not "eyed"). Identically distributed means that the chances of different outcomes are the same for all individual observations. If we're going to naively turn frequencies into probabilities, there's no way around some form of this assumption.[15] We are implicitly assuming all events are random. We are implicitly assuming that the order of our observations doesn't matter. We are implicitly assuming all events in isolation are the same. And, importantly, we are implicitly assuming that randomness in the future is the same as randomness in the past.

What do we get out of such assumptions in pattern recognition? Let's go back to my spreadsheet metaphor. Assume that every row in that sheet is an I sample from some distribution. Someone comes in and gives you two candidate functions that predict the special column from the other columns. The first function gets the predictions right for 80 percent of the rows, while the second function gets 90 percent of the predictions correct. What should we do with a new column generated by the same hypothetical random process that gave us the rows we've seen so far? Using our random logic, we'd put the chances that the first function gets the answer correct at 80% and the chances the second gets the answer correct at 90%. We'd clearly prefer the latter odds! This logic, based on the transference of randomness in the past to the future, lets us select the best model to run with. We now have an engineering dictum for machine learning. And we can now turn to the first time someone put this idea to use.

A Bell Labs in the late 1950s, Bill Highleyman was tasked with solving optical character recognition.[16] AT&T wanted to automate its billing system for long-distance calls. It's hard to believe now, but human operators connected long-distance calls in the United States well into the 1970s.[17] The operators wrote complex billing records for each call, and, in the '50s, these notes were transcribed by hand onto punch cards to be fed into computing machines for further charge processing. This sort of back-end human labor was inefficient, costly, and error-prone, and it seemed that the whole pipeline could be computerized. Highleyman was to build a machine that could automatically transcribe the operators' tickets onto punch cards.[18]

Highleyman wasn't quite sure how to design this machine from scratch. It was not clear at the time which part of the computations should be done with special-purpose hardware and which parts should be done with more general computers. There were all sorts of custom proposals for character scanning and no obvious choice for the best one. Highleyman and his colleague Louis Kamentsky thought that if they could capture the essence of written characters on a computer, the computer could *simulate* any scanning proposal and any pattern recognition scheme.[19] They hypothesized this dual layer of abstraction should make prototyping faster. This is one of those things that seems blatantly obvious in the present, but it was not obvious six years after the introduction of scientific computers.

Their first step was thus to design a general-purpose scanner to facilitate their simulations. This scanner would take an image of a digit and turn it into a 12 × 12 grid of pixels. This pixel grid would serve as the basis for the simulation of later OCR machines. Highleyman and Kamentsky could capture some representative digits and then use these digits to design the rest of their character recognition circuit.

The subsequent character recognition project would become Highleyman's PhD thesis.[20] This thesis outlines the core procedures of machine learning. He would define a problem of prediction, where the rows of the spreadsheet contained the pixel values of the image, and the special column was the character that this image corresponded to. He would explore different ways to *represent* the macro that mapped pixel values to character values. He would propose *optimization* procedures to find the prediction formulas that best fit his data. He would propose a split between *training* the mapping and *testing* the mapping. And he would propose creating a fixed dataset benchmark so researchers could compare their work. I don't want to claim that Highleyman invented machine learning. I just want to point out that these ideas are the most important concepts—I might go as far as arguing the only important concepts—in all of machine learning. They are all present in this document from 1961. Let's take a tour of Highleyman's work to see how his theory and practice of machine learning look identical to our own.

———

The first idea is to find a good function. Highleyman first restricted his attention to *linear predictors*. To explain linear prediction, let's return to my magic Excel spreadsheet and suppose the special column I'm trying to predict is binary. It can only take two values, whether they be 0 or 1, cat or dog, healthy or sick. A simple Excel formula I could try would be to take a weighted combination of the elements in my row, and if that combination is large enough, I'll label that row with a 1. Otherwise, I'll label it with a 0.[21] This sort of formula is a linear predictor.

To make this even more concrete, suppose we're trying to predict whether a person might soon have a heart attack. Row 1 equals 0 if the person has no history of heart problems and 1 if they have some history. Row 2 equals 0 if their EKG looks normal and 1 if it is abnormal. Row 3 equals 1 if they are over 50 years old and 0 otherwise. Row 4 equals 1 if they are a smoker and 0 otherwise. Row 5 equals 1 if their blood tests are abnormal and 0 if they are normal. Maybe we weigh these five different rows differently based on our clinical practice. If we multiply each number by these weights and up these rows, we'll get some score. If the score exceeds some threshold, we admit the patient to the hospital.[22]

Highleyman proposed linear classifiers as a reasonable means of distinguishing digits.[23] He could render the digits as pixels of a small grid. Each pixel would have a weight, and the class could be computed by adding up all the weights of the occupied pixels.

Highleyman wasn't the first to propose linear rules for classification problems. A few years earlier, Frank Rosenblatt had proposed a similar classification scheme that he called a *perceptron*.[24] For Rosenblatt, the perceptron was a model of a neuron that added up different weights and fired according to the presence of a strong enough weighting signal.

Though their formulations were mathematically identical, Highleyman had a hard time interpreting what Rosenblatt was after. He writes in a footnote how he's puzzled by why perceptron researchers don't use standard mathematical terminology. With regard to his classification rule, he states, "It goes by various names, such as artificial neuron, associative unit, and Adaline. In this paper, it will simply be called by its already well-established name of 'hyperplane.'"

Highleyman wasn't alone in his confusion. The language in the papers by the early researchers of artificial neural nets was strange to most engineers. It took a few years for the broader community to realize that Rosenblatt was not only fitting a simple linear predictor but was using similar techniques to what Highleyman would propose. In a seminal paper, mathematician Al Novikoff proved this equivalence.[25] Novikoff quipped that there was a "private language of perceptron workers" that obscured the simple core of their work.

Novikoff's translations showed that Highleyman and Rosenblatt both proposed optimization methods to iteratively improve a linear predictor. Assuming the random model of data generation, we'd want to pick the linear function that performs best on the collected data. But there are far too many possible linear functions to sort through exhaustively. For Highleyman's digits, there are 144 weights to assign. Even if each weight only took two different potential values, testing all combinations would be impossible. Both Rosenblatt and Highleyman proposed algorithms that found nearby functions that made better predictions on the dataset. Local improvement was achieved by the gradient descent algorithm. We discussed this method in the optimization chapter, but the high-level idea is pretty simple. Evaluate the current prediction function, and if it's making mistakes, nudge it a bit so it makes fewer mistakes.

Much to his credit, Rosenblatt's weight-finding procedure was remarkably simple. Let's say an image is supposed to be classified as positive, but the prediction is a negative number. We can describe the perceptron update at an individual pixel. If the misclassified image has a pixel value equal to v and the weight of that pixel is equal to w, then Rosenblatt's update sets the new weight value to $v + w$. He repeats this update for each pixel in the image. That's it.

Rosenblatt implemented this on an IBM 704 that could do about 12,000 arithmetic operations per second. His demo learned to distinguish between punch cards that were marked either on the left side or the right side. And yet, going from this update rule to what we do to train gigantic neural networks today only requires one extra step: computing the gradient efficiently. Rosenblatt's update rule is the backbone of more or less everything we do in contemporary machine learning.

Highleyman's algorithm was slightly more complicated but in the same spirit. Highleyman's perspective was to directly minimize the prediction error on the total sample of data at every step. His method computed the mistakes on the entire dataset and then looked for a slightly better model using a rule similar to Rosenblatt. Rosenblatt had focused on error-free rules, Highleyman focused on minimizing the probability of error, but our perspective from seventy years later suggests that they were basically doing the same thing.

As was evident in his method, Highleyman took a decidedly statistical approach to the problem of pattern recognition. He was explicit that this was the case. In his work about his random conceptualizations of data, he posed the problem by asking that we "consider the patterns to be randomly generated by a 'pattern source' according to the a priori probabilities of occurrence."[26]

Conceiving the data as random led to Highleyman's second foundational contribution: explaining why you need two sets of data to evaluate the quality of a pattern recognition machine. The first set would be used to find the best function, and the second would be used to verify that the machine's performance was up to snuff.

Now, recall that Highleyman's algorithm found a prediction function that minimized mistakes on a dataset. When data is random, the rate of prediction errors on the dataset should be similar to the rate of prediction errors on new data. Why would you need the second dataset? Why isn't the minimum error achieved through this process a good measure of the pattern recognition performance? There were multiple reasons. First, since the pattern recognizer is trained on a finite amount of data, it's plausible that more data might further improve recognition. Second, since it was tuned on a random set of data, it's possible that you would get an unlucky sample and have a machine that could have achieved higher performance if the data had been realized differently. Third, and perhaps most importantly, engineers had observed that prediction error rates on the data used to design pattern recognizers were often much lower than what they'd see when they'd demo their recognition machines to their colleagues and bosses. To avoid creating such undue optimism, new data was required to ensure more realistic error-rate estimates.

Highleyman showed two facts. First, if you just wanted to evaluate a single prediction function and not adjust its weights, you could get excellent error estimates using standard statistical tools. Second, Highleyman proposed a scheme to partition data between design and evaluation.

Highleyman asked what you should do if you had limited data. What would be the optimal way to partition the dataset into a set for model selection and a set for model evaluation? Highleyman derived a way to estimate the best possible error rate achievable. To get this best-case estimate, Highleyman computed that you should have at least as much data to evaluate as to fit parameters.

Highleyman's calculations turned out to be making a few too many idealizations and were probably wrong. But his idea of dataset partitioning stuck. Today, everyone partitions their data for design and evaluation. We call the dataset used to tune the model the *training set*. The data used to evaluate the model is called the *test set*. No one takes Highleyman's recommendation to make the training set smaller than the testing set. But no one has found a theoretical rule for what is best, either. Instead, practitioners lean on heuristics. The training set is always quite large. Sometimes, people partition data so that the training set is five times larger than the test set. Sometimes it's as much as 100 to 1. Generally, people aim for at least ten thousand examples in the test set. But, again, there are no definitive rules.

Alexey Chervonenkis, a pioneer in the mathematical theories behind pattern recognition, would recall that Highleyman's argument was intriguing but weird.[27] Chervonenkis credited Highleyman as one of the first people to propose the method of *empirical risk minimization* for machine learning. Empirical risk minimization is our technical term for choosing the model that minimizes some notion of average error on a training set. But Highleyman's statistical analysis didn't quite work out to the level of rigor Chervonenkis and his colleagues demanded.[28] Still, Chervonenkis and colleagues were intrigued. They figured out an appropriate solution that would lay the modern foundation for what we now call "statistical learning theory." Even though his calculations were wrong, Highleyman's ideas and practical methodology would pass the test of time.

Though empirical risk minimization and train-test partitioning are central, fundamental components of contemporary machine learning, Highleyman's most important contribution would be the introduction of dataset benchmarking, inspiring competitive testing that would rapidly advance the field of pattern recognition.

Highleyman's schemes for pattern recognition relied on data. What data could be better for character recognition than normal handwriting? Highleyman collected 50 alphabets from different people at Bell Labs. Each alphabet included all of the letters and the ten digits. Highleyman scanned these characters onto punch cards at a 12 × 12–pixel resolution. Example lists of digits from twelve of the writers are shown in figure 4. Figure 5 shows Highleyman's digitized scan of one of the full alphabets.

With the data of 1,800 alphanumeric characters in hand, Highleyman and Kamentsky could start exploring the different techniques proposed by other researchers for character

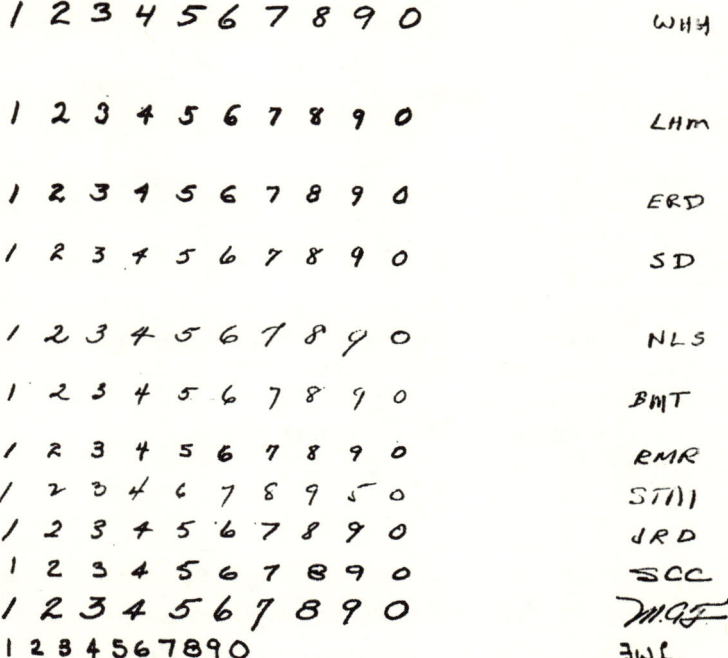

FIGURE 4. Examples of handwriting compiled by Highleyman to be scanned for early machine learning evaluations. This sample is taken from figure 32 of Wilbur Hull Highleyman's PhD thesis. "Linear Decision Functions, with Application to Pattern Recognition." Ph.D. dissertation, Polytechnic Institute of Brooklyn, 1961.

recognition. They began with a method devised by Woody Bledsoe from Sandia Labs. Highleyman and Kamentsky found that Bledsoe's accuracy numbers were not reproduced on their character dataset.[29] After Highleyman published this worrying suggestion, Bledsoe demanded an opportunity to defend his honor. Highleyman obliged, packaging a copy of the punch cards and shipping them across the country to New Mexico.

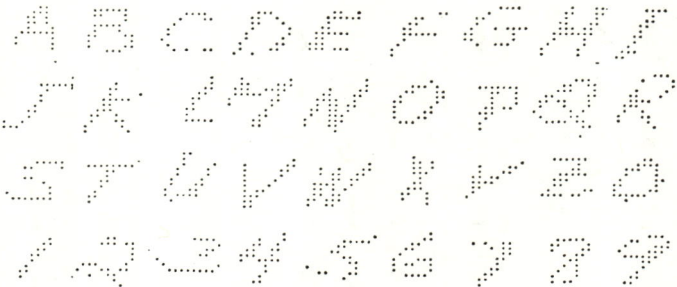

FIGURE 5. A sample of Highleyman's digitized alphabets. This display appeared as figure 7b in C. K. Chow, "A Recognition Method Using Neighbor Dependence." *IRE Transactions on Electronic Computers* EC-11, no. 5 (October 1962): 683–90. https://doi.org/10.1109/TEC.1962.5219431. Reprinted with permission from IEEE.

Upon receiving the punch cards, Bledsoe conducted his own experiment. In what may be the first implementation of Highleyman's train-test split, he divided the characters up, using 40 writers for training and 10 for testing. By tuning the hyperparameters, Bledsoe achieved approximately 60 percent error.[30] Bledsoe also suggested that the high error rates were to be expected, as Highleyman's data was too small. He predicted 1,000 alphabets would be needed for good performance.

By this point, Highleyman had also shared his data with Chao Kong "C. K." Chow at the Burroughs Corporation (a precursor to Unisys). A pioneer in using decision theory for pattern recognition,[31] Chow now turned to building a pattern recognition system for characters. Using the same train-test split as Bledsoe, Chow obtained an error rate of 41.7 percent,[32] using a more complicated prediction function that scanned the digit for certain patterns and then made a global decision based on the appearance of appropriate signals.

Highleyman's data created a buzz, and he began receiving frequent requests for his data. He thus decided to publicly offer to send a copy to anyone[33] willing to pay for the duplication and shipping fees. An interested party would simply have to mail him a request, and Highleyman would fire up the card duplicator and drop a box of 1800 punch cards in the mail. We have decidedly better systems to share data today. Based on citation surveys, Highleyman made at least another six copies of the data, with groups reporting results from Carnegie Mellon University,[34] Honeywell,[35] SUNY Stony Brook,[36] Imperial College,[37] The University of Wisconsin,[38] and Stanford Research Institute (SRI).

The SRI team of John Munson, Richard Duda, and Peter Hart performed some of the most extensive experiments.[39] They first used a method called "1-nearest neighbor": A card in the test set was given the label of the closest card in the training set. This method achieved an error rate of 47.5 percent. With a more sophisticated approach, they were able to do significantly better. Rather than working with pixel values, they created statistics about the individual images that they believed would be more discriminating. For each image, they scanned for line segments, counting their occurrences in each direction. Together, they ended up with 84 such statistics. They then made some combinations out of these 84 basic statistics. To fit weights to these resulting attributes, they used a version of Rosenblatt's perceptron algorithm developed by Carl Kesler.[40] With this more sophisticated approach, they were able to get a test error down to 31.7 percent, 10 percentage points better than Chow. When restricted only to digits, this method recorded 12 percent error. The authors concluded that they needed more data and that the error rates were "still far too high to be practical." They concluded that "larger and higher-quality datasets are needed for work aimed at achieving useful results." They suggested that such datasets "may contain hundreds, or even thousands, of samples in each class."

Munson, Duda, and Hart also performed informal experiments with humans to gauge the readability of Highleyman's characters. They found an average error rate of 15.7 percent on the full set of alphanumeric characters, about 2× better than their pattern recognition machine. Such low human accuracy suggested the data needed to be of higher quality. They concluded that "an array size of at least 20 × 20 is needed, with an optimum size of perhaps 30 × 30." As we will see, all of their predictions turned out to be spot on. We just needed to wait three decades to see this program through.

Let's reflect on the mechanism here. Highleyman knew data was needed to design a pattern recognizer. However, creating the data was labor intensive. It required the design of a special scanner and the recruitment of fifty volunteers. Rather than having every researcher create their own data, a shared dataset saved the community time. But it also created a *common task*. Instead of being at the whim of randomness, there was a deterministic task on which different research teams could evaluate each other. Comparisons became reproducible, and researchers could convince themselves that others evaluated their methods fairly. Probably more importantly, the competitive spirit of research teams inspired progress toward a focused end. Common tasks created competitions that could be won. Who doesn't want to show that their designs are cleverer than everyone else's? This paradigm of competitive testing on fixed benchmark datasets has driven all progress in machine learning over the subsequent fifty years.

But whatever happened to old Bill Highleyman? Since the computers in 1961 were so slow and the datasets so tiny,

Highleyman considered his doctoral project a failure. He (and Bell Labs) realized the computing of 1959 was not up to the task of character recognition.

> Although it was accepted as legitimate research by Brooklyn Polytechnic Institute, the thesis did not achieve its goals because of a lack of computer resources and reliable categorization algorithms.

After finishing his thesis, Bill abandoned pattern recognition and moved on to work on other computer engineering projects that he found more practical and engaging, never looking back. By the mid-sixties, Bill had immersed himself in data communication and transmission and patented novel approaches to electrolytic printing and financial transaction hardware. He eventually ended up specializing in high-reliability computing. Though he developed many of the machine learning techniques we use today, he was content to leave the field and work to advance general computing to catch up with his early ideas.

Highleyman's data is an anachronism decades ahead of its time. Indeed, despite the formation of journals and conferences, there was a great stagnation in pattern recognition methods after 1970. Papers from the 1970s in pattern recognition operated on tiny datasets that were seldom shared. By the time people started embracing the word "machine learning" in the 1980s, pattern recognition wasn't even part of the story. At this stage, Rosenblatt's work in the 1950s had been dismissed as a failure, but AI researchers like John McCarthy and Herbert Simon still thought that a learning machine should be possible, even though no one was quite sure what the word "learning" really meant. There was a camp that founded the International Conference on Machine Learning (ICML), dedicated to techniques from logic and search heuristics (what some now call "good old-fashioned artificial intelligence").[41] Another camp was more inspired by neuroscience, founding the Advances in Neural Information Processing Systems (NeurIPS) conference.[42] This group aspired to provide an entirely new model for computation itself, with hardware and software designed to mimic the brain.[43]

The problem is, none of this stuff really worked. Machine learning didn't fully become a field until it rediscovered pattern recognition and competitive testing in the late 1980s. What was special about the late 1980s? First, email and file transfer were becoming more accessible. The current specification of the standard protocol for transferring files on the internet, aptly named File Transfer Protocol (FTP), was finalized in 1985. In 1987, a PhD student at UC Irvine named David Aha built an FTP server to host datasets for empirically testing machine learning methods. Aha was motivated by service to the community, but he also wanted to show that his nearest neighbor methods would outperform Ross Quinlan's decision tree induction algorithms. He formatted his datasets using the "attribute-value" representation that Quinlan had adopted with his TDIDT (top-down induction of decision trees) algorithm.[44] The UC Irvine Machine Learning Repository[45] was born.

The other notable shift in machine learning was a demand from funding agencies for more quantitative metrics. AI had found itself in one of its perennial funding winters, and program managers demanded more "results" before they'd be willing to write grant checks. In 1986, DARPA (Defense Advanced Research Projects Agency) program manager Charles Wayne proposed a speech recognition challenge where teams would receive a training set of spoken sentences and be evaluated by the word error rate their methods achieved on a hidden test set.

Wayne worked with the National Institute of Standards and Technology to create and curate this dataset, which we now know as TIMIT (named after the labs that helped to put it together at Texas Instruments and MIT). TIMIT was too large to share via file transfer, so NIST released it on CD-ROM, the punch card of the 1990s.

Improvements in computing greased the wheels, giving us faster computers, faster data transfer, and smaller storage footprints. Demand for quantitative metrics forced us to find consistent, reliable quantities for comparison.

But no matter how hard I try to come up with an explanation, I can't understand how the only sticky part of machine learning is this core of mindless pattern recognition. How is it that the shift in the 1980s ends up looking exactly like the paradigm Highleyman kickstarted in 1959? How is it that early ICML and NIPS had no idea that pattern recognition would be the only thing that would work? How is it that for every "benchmark" task that comes out, we end up with a train-test split and some simple, climbable metric? And why is it that certain benchmarks carry more weight than others? I don't have answers, but the next three decades would bring a string of increasingly more impressive examples.

In 1994, once again at Bell Labs, researchers released a dataset of handwritten characters. The goal was to showcase the massive improvements in pattern recognition in the few short years since its revival in the 1980s.[46] Comprised of a dataset of digits from the National Institute of Standards and Technology, the MNIST dataset consisted of 70,000 digits. There were no letters in this dataset. The digits were written by 500 high school students and some workers at the US Census Bureau. The writers wrote multiple versions of the numbers so that the data could capture intrapersonal variability. The digits were represented at 28 × 28 resolution, more than twice what Highleyman used. The images were rendered in grayscale rather than simple binary values, preventing certain visual artifacts. The dataset was split into 60,000 training examples and 10,000 testing examples.

With this doubled resolution and a training set 150 times larger than Highleyman's, a neural network achieved 0.7 percent test error on this new digit dataset. In fact, a similar model to Munson's architecture that uses combinations of edge detectors achieves a similar 0.6 percent error rate. The 1960s intuition of Bledsoe, Duda, Hart, and Munson proved correct. They had called for a doubling in resolution and a 100-fold increase in dataset size. Lo and behold, this sufficed to solve handwritten digit recognition.[47]

Unlike Highleyman's data, MNIST featured only digits, no letters. Only recently, in 2017, researchers from Western Sydney University extracted alphanumeric characters from the National Institute of Standards and Technology data repositories.[48] The resulting *EMNIST_Balanced* dataset has 2,400 examples in each of the 47 classes, with a class for all uppercase letters, all digits, and some of the non-ambiguous lowercase letters. Currently, the best predictor has a 9 percent error rate.[49] This is only a 3× improvement over the methods of Munson, Duda, and Hart.[50] Considering that the SRI team observed a human error rate of 11 percent on Highleyman's data, an accuracy of 90 percent may be close to the best we can expect for recognizing handwritten letters and digits without additional context.

This evidence shows how spot on dataset benchmarking can be. Thirty years later, the predictions from the 1960s have more or less borne out, and the approaches developed then only needed modest refinements on the larger data. As we shall now see, competitive testing on a particularly compelling dataset ushered in our contemporary age of artificial intelligence.

The perhaps most convincing evidence of the power of the dataset would come in 2012. Let me begin in 2009. Researchers at Princeton led by Fei-Fei Li compiled a massive database of images from the internet called ImageNet.[51] The team started with another Princeton database called WordNet.[52] WordNet collects words into "synsets" of words that express a single concept. The synsets were connected by subordinate relations so that "bass" and "trout" would be connected below "fish." By searching the photo-sharing website Flickr, Li's team tried to find 500–1,000 images for each of the approximately 80,000 synsets in WordNet corresponding to nouns. To do this, they would first search Flickr[53] for candidate images and then, through a massive crowdsourcing effort, throw out the photos that didn't match the term concept.

The images were annotated by hiring workers through Amazon's Mechanical Turk service.[54] Workers were shown a collection of images and asked which contained the Flickr search query. One of my favorite examples is finding images containing a "bow." When you type "bow" into an image search engine, you'll find images of weapons, ribbons for presents, and people showing deference. But if I tell you that I mean "a weapon for shooting arrows, composed of a curved piece of resilient wood with a taut cord to propel the arrow," you can quickly discard the erroneous searches. This was the task given to the Amazon MTurk workers.

Shortly after assembling this dataset, Li and her collaborators decided to host a new challenge in pattern recognition.[55] At the time, most popular datasets for pattern recognition in images had tens of thousands of data points. The larger ones only had low-resolution images. What would happen if the same challenge was hosted with millions of images, thousands of categories, and the resolution of images posted on Flickr? The ImageNet team assembled a subset of their images for the competition. They chose a set of 1,000 categories of different images from their synsets. Of these, 118 were breeds of dogs introduced so that researchers could understand the fine-grained abilities of different pattern recognizers. Most other classes included higher-level concepts like "compass," "racket," "canoe," "mug," "tank," "bell," and "airliner." The compiled training set had over 1.2 million images, with around 1,000 images per category. The testing set had ten examples from every category.

The ImageNet competition was pretty niche for its first three years. By point of comparison, Netflix had run a competition to improve their internal recommendation systems a few years earlier, and thousands of teams competed in the first year alone.[56] Since the Netflix dataset could be analyzed on a laptop, the competition was widely accessible (though Netflix offering a $1 million prize to the winning team also amplified interest). By contrast, entering the ImageNet competition required processing millions of high-resolution images. Only committed teams could enter, and there were only nine teams in 2010, four in 2011, and six in 2012. Despite the sparse participation, the competition drove progress and led to the most important finding in the last two decades of machine learning.

In 2012, a new team from the University of Toronto entered the competition and outperformed everyone. The Toronto team's method provided correct answers on 85 percent of the images. The next best team was only accurate 75 percent of the time.[57]

Whereas other teams built sophisticated features from the data, this team followed the tried and true path of Highleyman. They proposed a particular form for the function that would take an image and compute a prediction. They searched for the image that produced the fewest mistakes using gradient descent. Certainly, their functional form was far more complex than Highleyman's linear predictors. They used what is called a "convolutional neural network," which, though inspired by the brain, is simply a convenient way to write down functions that combine spatial information in images. And they had to go to great lengths to process all of the data. Lead author Alex Krizhevsky wrote custom code to efficiently network two graphics cards together, stitching together a DIY supercomputer that could optimize the weights in his model in about six days.

The Toronto team's competition success was undeniable, leading to an industrial revolution in neural networks. The next decade saw frenzied interest, the production of fast, open source code to train these prediction systems, and intense competition to improve upon each other's results. Neural network models would soon label images on social media, translate between different languages, and power large language models. There is no theory of why "deep neural networks"—as we now call them—work well on all of these different prediction problems. Their consistent success in competitive testing sustains them as cutting-edge technology. Since we have found nothing better, the program of proposing a complex functional representation, using a dataset to evaluate the best fit, and finding models with gradient descent is now the standard approach in all of machine learning.

At a workshop in 1980 that would become the annual International Conference on Machine Learning (ICML), Herbert Simon delivered a plenary about the past, present, and future of machine learning. Given that computer power had exponentially exploded in the twenty-two years since Rosenblatt's perceptron, why would we build machine learning systems at all?

> If one thinks about [it] a little, one says . . . "Who—what madman—would put a computer through twenty years of hard labor to make a cognitive scientist or a computer scientist out of it? Let's forget this nonsense—just program it." It would appear that, now that we have computers, the whole topic of learning has become just one grand irrelevancy—for computer science.[58]

Simon went on to survey some of the successes of classical machine learning and how they were disappointing. All his examples were from the 1950s, when we were stuck programming on vacuum-tubed mainframes: Samuel's checkers player, one of Simon's own pet projects for pattern recognition called EPAM trees, and the perceptron. With regards to the perceptron, Simon didn't mince words.

> A final "classical" example (this is a negative example to prove my point) is the whole line of Perceptron research and

nerve net learning [Rosenblatt, 1958]. A Perceptron is a system for classifying objects (that is, a discovery and learning system) that computes features of the stimulus display, then attempts to discriminate among different classes of displays by computing linear additive functions of these features. Functions producing correct choices are reinforced (receive increased weight), those producing incorrect choices have their weights reduced. I have to conclude (and here I don't think I am in the minority) that this line of research didn't get anywhere. The discovery task was just so horrendous for those systems that they never learned anything that people didn't already know. So they should again strengthen our skepticism that the problems of AI are to be solved solely by building learning systems.

I wonder what he'd think of ICML 2025, which is overwhelmingly perceptron research.

There are two knee-jerk reactions to Simon's quote. The first is, "What a fool! Perceptrons are everywhere, and it was only after embracing the perceptron that machine learning took over all of computer science and engineering." The other reaction is, "What a genius! Perceptrons are everywhere, and machine learning systems are inefficient monsters that don't really work even though we've thrown billions of dollars at them." There is truth in both statements.

It is certainly undeniable that Simon's hyperbole was fighting fire with fire. Artificial intelligence has always been plagued by arrogance, hyperbole, and myopia. Even the name is a marketing term. It's hard to blame Simon for this dismissal when, only twenty-two years earlier, the lede of a *The New York Times* article declared that Rosenblatt's perceptron would become conscious of its existence.[59]

Rosenblatt thought that his simulated neuron would become conscious of its existence. Simon thought that machines were decidedly different from people. How do we reconcile these two visions? To find that middle ground, let me bring up Marvin Minsky and Seymour Papert's attack on the perceptron.

In their 1968 book *Perceptrons*,[60] Minsky and Papert famously showed that the perceptron couldn't learn the *parity function*. Using rigorous mathematics, they showed that a perceptron couldn't learn to predict whether the number of 1s in a bit string was odd or even. But let me tell you why this is a positive case for machine learning.

Think about the sorts of pattern classification tasks where machine learning works well. For example, part of the ImageNet classification task is distinguishing between 118 dog breeds. If you show a trained human one of these pictures, they can learn to distinguish them with over 80 percent accuracy.[61] Artificial neural net models can do slightly better than this. But we don't know how to write a computer program with competitive dog breed accuracy that doesn't use machine learning.

The standard resolution of a picture on an iPhone 14 is 4 megapixels, about 100 megabits when uncompressed. A human looking at an image of this size can classify a dog. But imagine showing someone a bit string with 100 million ones and zeros and asking whether the number of ones was even or odd. Not a single person could do this. And yet, I can compute the parity of a bit string by evaluating the Python code:

$$X.\text{sum}() \% 2$$

This calculation takes milliseconds on a laptop. I can write a trivial program to compute the parity of the bits representing a file, but I can't *see* parity.

People can learn to make highly accurate predictions for dog breeds, but writing a syntactically correct computer program to complete this task feels impossible. No human can see parity, but writing a computer program to compute parity is beyond trivial. This paradox drives artificial intelligence researchers insane.

People read Minsky and Papert's book as claiming that perceptrons should be abandoned. However, in their revised edition, which appeared twenty years after the initial publication, they emphatically argued that this wasn't what they were after. Minsky and Papert wanted an understanding of knowledge representation. What was the right way to arrange the bits so that computers could be guaranteed to recognize patterns? They wanted a mathematical definition of what makes a pattern recognizable.

Unfortunately, the best answer seems to be that if someone can get their machine learning code to work, then the pattern is predictable. Conversely, if someone can prove a pattern is not predictable, I can write a short computer program that gets 100 percent accuracy.

In some sense, this shouldn't be surprising. If I have a simple way of *mathematically describing* how to predict B from A, that implies I have a way of *writing code* to predict B from A. What is code if not a simple mathematical description? Machine learning only makes sense if an engineer *doesn't know* how to write the code. Instead, the engineer rests their design on the belief that prediction is possible and that they can collect data of the pattern they are hoping to associate.

I believe handwritten characters correspond to letters in the dictionary. I believe that animals can be identified in pictures. I believe that heart disease could be diagnosed from a few simple tests. I believe that protein structure can be predicted from amino acid sequences. If an engineer shares any of these beliefs, they can use machine learning to search for a function that computes the associated predictions. They just need to collect enough data and access sufficient computational resources.

Here's my quiz to determine whether a pattern classification problem is solvable.

1. Does conventional wisdom say there should be a classification rule that is stable over time?
2. Can we not write a computer program that does it?
3. Can we collect a lot of data?
4. Are the stakes low if you make an error?
5. Do you have access to a ton of computing power?

If you answered yes to all of these questions, then the problem can likely be solved by machine learning. Machine learning is for problems where we are convinced there is a good prediction rule but can't articulate how to write this rule in computer code.

There are interesting, important problems in this space. Notably, one I just mentioned: predicting the structure of a protein from its amino acid sequence. Machine learning techniques led to the biggest breakthrough in protein structure prediction in decades.[62] Using a competition model, machine learning–based approaches were able to "solve" protein structure prediction. This work, Google's DeepMind's "AlphaFold," has been so impactful in structural biology that it was awarded the Nobel Prize in Chemistry in 2024.[63] We should be dazzled that it works as well as it does.

And to come back to them one last time, large language models like ChatGPT are nothing but the Shannon-Highleyman project writ large. Yes, complex functions compute the probabilities, but the models are trained using gradient descent to minimize the average prediction error on a dataset. It's just that now the dataset is all text ever produced by humankind. That these models can create realistic chatbots that can convincingly do all sorts of text summarization, boilerplate generation, and information retrieval is remarkable. We have long believed language is predictable, but the scope of how language models replicate language exceeds anyone's predictions.

Make a model large enough. Make your data large enough. Add enough layers. You'll find a perfect prediction rule eventually. Or you won't. Either way, the only way you'll know is if you try. Machine learning is what we do when we don't understand. When we do understand, we can just write code.

I can understand why this state of affairs is disappointing to people who want formal reasons for why technology works. Our quest to make it rigorous has been a mixed bag. I worry that by adding formalism, we are just trying to convince ourselves we're doing the right thing. Too often, formalism causes us to forget we have no idea what we're doing.

Though we don't have particularly informative theory, we do have a well-defined core for when and how machine learning works. We need a clearly defined goal for the task. We need a good dataset and a way of turning it into a spreadsheet. We split this dataset into training and testing sets. We need reasonable means to represent prediction functions on the data. Then, and only then, we run gradient descent to build a prediction machine. But after this point, the machine learning stack is standard enough for programmers to get up and running immediately with open-source software.

Getting all the parts to align is an art that requires practice and skill. Sometimes our faith in machine learning methods is misplaced. Notably, machine learning models in medicine have failed to deliver despite a good deal of optimism. After winning

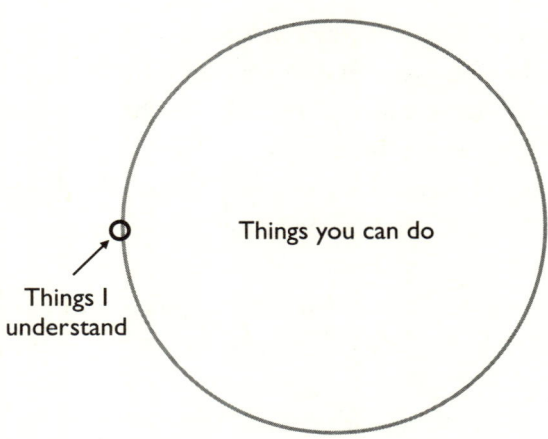

FIGURE 6. A Venn diagram of possibility versus understanding.

on *Jeopardy*, IBM Watson devoted teams to scaling up machine learning healthcare applications, only to shut down after repeated failure in 2002.[64] Epic Systems, the largest vendor of electronic health records platforms in the US, shipped tools for detecting the onset of sepsis that doctors only found to be wholly unreliable after performing an extensive validation in their hospital.[65] Machine Learning for Healthcare is not yet ready for prime time. As suggested by question 4 in the quiz above, we should expect machine learning models to make mistakes. The stakes associated with such mistakes must be low. I'll dig more into how we might properly weave statistical predictions into human decision making in the next chapter.

Despite negative results, we can't deny that machine learning methods have done thoroughly mind-blowing things. AlphaFold transformed our approach to protein structure prediction, and ChatGPT has transformed how we interact with computers. Beneath the scenes on your phone, machine learning touches up your photos and manages your social media timelines. What else might it be able to do in the future? Figure 6 sums up our current understanding of pattern recognition and machine learning.

What I describe in this chapter is most of the tiny circle and a little bit of the big one. And maybe that's the most important takeaway: To push the envelope in machine learning, you have to get yourself into that blue circle while trying to expand the red. To do the mind-blowing, you have to work with that which you don't understand.

6

Humans Against the Machine

ACROSS THE last four chapters, we've seen how mathematical rationality and its associated computational decision making was conceived to improve upon the decisions of people. Optimization was created to make planning less ad hoc. Game theory envisioned a world in which we could make optimal decisions by strategizing against an imagined adversary. Randomized experiments aimed to remove clinician biases in treatment evaluations. Statistical machine learning hoped to automate human data processing, pattern matching, and prediction. The history of these fields revealed how each of these pillars is remarkably powerful in the appropriate contexts but limited as a universal basis of decision making.

Despite these limitations, can we rule out that mathematical rationality might still be preferable to unfettered human decision making? People may very well just be flawed computational decision makers. Perhaps mathematical rationality, even with all its imperfections, can produce a better result than a human. Motivated by this possibility, and in parallel to the development of computers as the ideal rational agent, psychologists, economists, and even computer scientists set out to understand the ways in which humans *were not* computers.

Merrill Flood and Melvin Dresher's experiments on the Prisoner's Dilemma at RAND, discussed in the context of game theory in chapter 3,[1] were some of the first such investigations. By the early 1950s, they had already found that humans and mathematically rational machines strategized in completely different ways. Flood and Dresher's experiments showed convincingly that humans did not play games *optimally*, and hence weren't mathematically rational. There are two ways to interpret this finding. The first is to take this as a knock against humans and to investigate the many ways that people are "irrational." The other perspective asks, well, then what kind of rationality *do* people use? These two different interpretations mirror two camps of adversarial research that followed Flood and Dresher's work, each with their own way of explaining human decision.

One community of researchers began with the opinion that human deviations from mathematical rationality were bad and undesirable. In a series of experiments, they found that not only would humans not abide by the rules of mathematical rationality, but they would predictably deviate from rational choices in systematic ways. This predictable irrationality is the core of a very influential school typically associated with *behavioral economics*. This field aims to see how people make decisions when boxed into the models economists use to think about their markets. I'll use a name they deemed appropriate for themselves, the "Heuristics and Biases" (HB) school.

The second community tried to understand how people were able to think outside the confines of the mathematically rational box. They wanted to understand what made people different from machines. They investigated how trained experts could make time-critical, lifesaving decisions in the flash of an eye, often finding solutions that no one could have conceived of beforehand. This community sought to determine what made people special and how to empower human decision makers who engage in high-stress, high-risk tasks. I'll refer to this second group as the "Naturalistic Decision Making" community (NDM).

To their credit, each group discovered valuable interpretations of how people make decisions. The HB community focused on the hazards of human decision making, whereas the NDM community focused on what made human experts indispensable. This chapter chronicles these two adversarial research camps of research, explaining how each viewed human decision making in light of advances in automated decision making. Let me start by describing a bit about the two schools, how they differ, and where they surprisingly agree.

Throughout his undergraduate and graduate studies in Minnesota, psychologist Paul Meehl found himself at the center of a personal and professional conflict for the soul of psychology. As a high schooler, Meehl had been drawn to psychology by the psychodynamic school (pioneered by Freud), which considered the myriad connections between a patient's past experiences—even their dreams—and their current mental state. For psychodynamicists, people were a complex manifestation of their past experiences, and it required a skilled clinician to uncover the latent trauma that caused their current distress. At Minnesota, Meehl was educated by a rigid behaviorist crew that was strongly anti-Freudian. The behaviorists focused on understanding the impact of external factors on mental states. They were adamant that behavior could be understood through rigorous, scientific tabulation of statistical data. Meehl remained passionate about both hard scientific evidence and philosophy. He practiced clinical Freudian psychoanalysis but also ran rigorous experimental lab work on behavior in mice. He effortlessly moved between two juxtaposed worlds: the world of the hard quantification and the world of the metaphorical and unquantifiable.[2]

Meehl's career, starting as an assistant professor in 1945, spanned the computer age. It is thus no coincidence that his most influential work, his 1954 book *Clinical versus Statistical Prediction*,[3] set out to determine whether people or machines made better decisions.[4] Even as computers were just coming online, there was already ample evidence that statistical pattern recognition could, and perhaps should, play a role in critical decision making. Though he wouldn't call it by name, Meehl's book makes the first argument for machine learning in the clinic.

Meehl's interest was in making consequential decisions about human outcomes, especially in the context of clinical psychology. Should a person be admitted into a school given their test scores? Should a prisoner be released on parole given their record of behavior? Should a patient suffering from clinical depression be hospitalized to prevent suicide? Should a person who doesn't respond to antidepressants receive shock therapy?

Meehl set out to compare decisions based on statistical prediction to those based on the judgment of a doctor or psychologist (clinical prediction). The simplest form of statistical prediction was the *actuarial method*, where, as in the last chapter, rates of past success were used to predict future outcomes. In the actuarial method, each patient would be assigned to some category, and the prediction for that patient would be the average outcome of all patients of the same category in the past. For example, you might predict the likelihood a person does well in school by the average of all previous students with similar test scores. Meehl generalized the actuarial method to his notion of statistical prediction, which meant any mechanical procedure that took as input a set of data and output a predicted outcome based on past observations. This is the same general notion of statistical prediction I used in the last chapter, and today is what we broadly refer to as machine learning.

By contrast, Meehl never cleanly defines clinical prediction. It refers to any "informal" decision making made by a human specialist. Meehl used this term broadly to capture whatever process occurs in people's heads. Clinical rules are those based on intuitive assessments of data by clinicians and are not apparently formalizable as algorithms.

Meehl found there was a particular sweet spot for statistical prediction. For problems with a small set of outcomes and actions, a clean collection of measured data, and considerable uncertainty about the result, statistical prediction provided more accurate judgments about the future than clinical judgment.

Meehl detailed several examples of where statistical tables seemed to produce better predictions about outcomes than clinical psychologists. He cited the work of sociologist Ernest Burgess, who worked in the 1920s with the Illinois Parole Board to determine the factors that contributed to recidivism. Burgess showed that statistical methods could predict whether a parolee would commit further crimes after release. His simple checklist of 21 predictive factors was more accurate in predicting recidivism than two prison psychiatrists.[5] Burgess argued that mechanical systems were potentially fairer than human judges as they could accurately predict who was at risk of returning to crime while being unprejudiced by politics.

Meehl highlighted a dozen other studies in his book, from predicting performance in college to remission of schizophrenia, and continued to track these throughout his career. No matter how much he looked, he kept finding the same thing as Burgess: Statistical rules were seldom worse and often much better than clinical predictions. In a reflection on his book, Meehl wrote in 1986, "There is no controversy in social science that shows such a large body of qualitatively diverse studies coming out so uniformly in the same direction as this one."[6] Two meta-analyses on the topic in the 2000s further confirmed Meehl's findings.[7]

Many have taken Meehl's argument to suggest that statistical prediction is always better than clinical judgment in all decisions. This is a misreading. Meehl identified a narrow but widely applicable sweet spot. To use statistical prediction, there needed to be clear, cleanly defined outcomes. The problem needed to be answerable by multiple choice. There needed to be a clear set of data that could be ingested by a computer. There needed to be a set of reference cases to compare each case to. Only with these ingredients could a statistical model be built. However, in these cases, empirical evidence over decades showed that a statistical model would perform as well or better than human experts.

But why were humans not better than machines in this case? This question inspired what would become one of the most influential research programs in economic psychology of the last fifty years. Meehl's 1954 book, along with Flood and Dresher's work at RAND, remains the bedrock of HB research. It would define an influential school arguing in favor of mechanical decision making. A school that would argue that humans are predictably irrational.

From 1955 through 1958, Daniel Kahneman served as a psychologist in the Israel Defense Forces. One of his assignments was assessing whether recruits should be enrolled in officer training. Much to his surprise, Kahneman found that his "ability to predict performance at the school was negligible."[8] This was a prediction problem precisely of the character described by Meehl: The outcome was simple to measure (making it through officer training), the entrance examinations presented a clean collection of data, and there was always uncertainty about any candidate's outcome. Meehl's book helped Kahneman frame his predictive difficulties. His unexpectedly poor predictive performance and Meehl's book inspired him to build a standardized testing protocol for predicting a candidate's success in officer school. Though the IDF interviewers initially balked at using the test and "being reduced to unthinking robots," the results would prove far more predictive than the clinical expertise. Kahneman's screening system was adopted and would remain an integral part of the IDF officer screening process for decades.

After his military service, Kahneman got his PhD in psychology and became a professor at the Hebrew University of Jerusalem. In 1968, he began collaborating with his colleague Amos Tversky on the foundations of human decision making. Through a series of clever experiments in the 1970s, Kahneman and Tversky showed decisively that humans were are not mathematically rational.

Kahneman's military experiences inspired his work with Tversky some fifteen years later. Officially founding the HB school, Kahneman and Tversky would precisely hone in on how human decision makers had consistent and predictable *heuristics and biases* that caused them to deviate from statistical optimality. The biases discovered by Kahneman and Tversky include the *availability heuristic*, where people guess probabilities of events based on the ease of examples coming to mind as opposed to by careful counting of the correct frequencies. One of their early experiments asked people to guess whether the letter R appeared more frequently as the first or third letter in a word. Most people answered the first position, but the actual answer was third. Kahneman and Tversky observed that people would bring to mind the many words they could think of containing the letter R and make a statistical judgment based on this false, mentally simulated rate.

Kahneman and Tversky constructed dozens of similar experiments and found countless instances of false confidence. In addition to the availability heuristic, Kahneman and Tversky found dozens of other irrational heuristics and biases. They discovered the *anchoring heuristic*, where people estimate likelihoods based on irrelevant information seeded by the questioner. They found the *representativeness heuristic*, where people incorrectly associate some outcome by pattern matching to examples they've seen before. All of these cases seemed to explain *why* clinical judgment was often suboptimal: because people would fall prey to the biases inherent to their very nature. To maximize accuracy, you needed to work like an actuarial table. The heuristics people used introduced statistical biases that distorted their predictions.

There are two different ways to interpret these sorts of HB experiments. The first is that they show something wrong with people. Since people deviate from the artificial formulation of mathematical rationality, this means they are not looking out for their best interests. Most people who advocate in favor of the HB school of thought take this viewpoint. I reject this interpretation, and will discuss why later in the chapter.

But first, I want to present a second, more positive way of interpreting Kahneman and Tversky's important work as it aligns with the evidence I've been compiling so far in the book. Let's accept the argument I've made that mathematical rationality can't solve all decision-making problems, but it shines inside the narrow regimes where it is applicable. If we could delineate when and why people are less reliable than machines, we could envision systems that guide people to make better decisions. I'll return to this idea in the concluding chapter: How can mathematical rationality be most usefully applied to the future of human decision making?

Paul Meehl provided one such setting where mechanical decision making is probably better than clinical: answering clear, multiple-choice questions about simple actions from machine-readable data. Because of the biases identified by the work started by Kahneman and Tversky, computers will likely always have an edge in this particular mode of decision making. Kahneman's predicting performance in officer school is an archetypal example. However, what "machine-readable" means is a moving target, and a function of data quantity and computing speed. Today's faster computers and mature software can tackle considerably more complex decision problems.

Meehl's characterization is helpful because statistical rules are more accurate, faster, and cheaper than experts. They can potentially be more fair and safe. If we can use statistical prediction or machine learning, we can improve our decisions. If you accept this argument, it seems reasonable to argue that not only should we use algorithms for such decisions, but we should train ourselves to act more like algorithms when making predictions. That is, we should implement policies that save people from their predictable irrationality.

Whether or not you agree, it's undeniable that the more influential interpretation of Kahneman and Tversky's work with economists, policymakers, and public pundits was the one that assumes people's irrationality needs to be corrected. If we accept that humans predictably err, perhaps it's worthwhile to guide humans to be more like rational robots. Mechanical rules, even non-statistical ones, could potentially improve human processes.

A particularly successful application of this idea is safety checklists. Such checklists can help ensure a common starting place for dangerous activities. For example, Peter Pronovost introduced safety checklists for intensive care procedures in hospitals. His most famous checklist involved cleanliness around the insertion of a deep line into a vein during surgery. Nurses were granted authority to correct colleagues—including doctors—if they violated procedures. Infections from venous lines dropped from 3 per thousand to 0. That's right: zero. Just by including a five-step mechanical checklist, infections due to this particular procedure were effectively eradicated.[9]

Checklists, rules, and mechanical guidelines now rule much of the medical decision making in the United States. For the hospital administrator, there are clear advantages to forcing decisions into simple contexts. In addition to minimizing accidents, there is ample evidence following Meehl's work that statistical rules make better, more cost-effective care decisions than clinicians in a variety of contexts. Though there are few examples as cut-and-dried as Pronovost's infection control protocols, consistency of care has been consistently proven to improve the average outcomes in hospitals. Healthcare should aim to deliver benefits to as many as possible, and one way to do this is by improving the mean outcomes. It is hard to argue against the idea that the standard of care can be lifted by removing biases, oversights, and errors by healthcare workers. No one wants to argue against improving the standard of care.

This idea applies far beyond healthcare. Automated decision rules can improve standards in many contexts, with a long and notable legacy in airplane safety. They also can make goals clear. Automated rules necessarily chase clear, well-coded outcomes, and tracking these outcomes helps management track and find quantitative improvements to their services. Moreover, rules can accelerate human decision making. Well-designed rules let us not worry about every minute detail that can arise in stressful environments.

And though many books are written to claim otherwise, mechanical rules can often be fairer than human decision makers. This is especially true if the rules are designed with appropriate consultation of stakeholders and an agreement on how fairness will be evaluated. Rule-based systems are free from ambiguity and human unpredictability. This objectivity and reliability of mechanical rules can instill a broad confidence in their effectiveness.

Psychologist Gary Klein, one of the central figures in the NDM school of thought, spent his career studying how experts made decisions in high-stakes environments and how to build tools to better enable such human expertise. In his book *Streetlights and Shadows*, he lauds the success of checklists like those proposed by Pronovost as correctives for some of the issues raised by HB researchers.[10] He notes that similar checklists have led to dramatic safety improvements in aviation as well.

However, Klein cautions that checklists are not a panacea. There are many downsides that we have to consider in mission critical environments like aviation and healthcare. Mechanical rules and checklists are challenging to keep updated. It is difficult to change standards once they are in place, and if an agile solution is required for a dynamic world, checklists will stand in the way of adapting.

The same complaints can be levied at statistical decision aids. Statistical predictions can be used similarly to checklists, surmising from the current data which actions are most likely to yield successful outcomes. Dozens of statistical decision aids are used in healthcare practice, helping physicians assess serious conditions, from whether ICU patients may be progressing into septic shock to whether ER patients are experiencing major adverse cardiac events. But even if we accept Meehl's premise that statistical prediction is never worse and often better than clinical judgment, and Kahneman and Tversky's argument that people are predictably irrational, that also doesn't mean that statistical rules are infallible. Accurate statistical prediction remains a delicate skill. Incorporating the appropriate information in a proper machine-readable format is an art form. And past results must always be indicative of future outcomes.

One thing proponents of the HB school often fail to mention is that there remain *simple* decision problems where machines fail, but humans succeed. Computers are predictably irrational in their own way. Deceptively simple automation tasks can't be formulated as optimizing in the face of uncertainty. My current favorite example of something most unskilled humans can do but machines have still not mastered is driving a car.

Self-driving cars have been envisioned as a possibility since the 1950s.[11] For example, a 1957 advertisement by America's Independent Electric Light and Power Companies declared,

> One day your car may speed along an electric super-highway, its speed and steering automatically controlled by electronic devices embedded in the road. Highways will be made

safe—by electricity! No traffic jam ... no collisions ... no driver fatigue.

A 1958 Disneyland TV program, "Magic Highways," narrated:

As father chooses the route in advance on a push-button selector, electronics take over complete control. Progress can be accurately checked on a synchronized scanning map. With no driving responsibility, the family relaxes together. En route business conferences are conducted by television.

Since every adult you know can do it, it seems like automating driving should be relatively easy. The goal is to get from point A to point B as quickly as possible while obeying traffic laws and not hitting anything. This broader task appears like it can be broken down into subproblems perfectly suited for statistics and optimization. Find shortest routes using linear programming. Put a camera on the front of the car and set up an optimization problem to find a lane. Build a control system to track the center of the lane and follow the speed limit. Use a radar system to detect objects in front of you and program your control system to stop if something is in the way. Model other drivers as adversarial agents and compute game theoretic strategies to interact with them.

Researchers at Carnegie Mellon University built demos of such capability in 1988, showing that approaching automatic self-driving like this was promising.[12] However, computers in 1988 were still bulky and slow. The CMU project vehicle was a giant van packed full of clunky computers, monitors, and sensors. It would be another sixteen years (and eight doubling cycles of Moore's Law) until computers were fast enough and small enough to really make self-driving appear tractable. In 2004, the US Defense Advanced Research Projects Agency

(DARPA) held a contest to see who could build such a self-driving system to navigate a 150-mile course through the California desert. Though none of the entrants finished the route, the cars had come a long way, sparking broad interest among both the military and technologists. According to DARPA's reporting: "These challenges helped to create a mindset and research community that a decade later would render fleets of autonomous cars and other ground vehicles a near certainty for the first quarter of the 21st century."[13] In 2009, Google funded a self-driving car project led by Stanford's Sebastien Thrun, whose team completed and won the second version of DARPA's grand challenge race in 2005.

Funded by Google's massive coffers and founders' desires to build flashy "moon shots," self-driving car research and design flourished. By the early 2010s, technologists had convinced themselves that the cars were inevitable. In 2012, three years into their research program, Google released a famous ad showing their self-driving car driving a legally blind man around Silicon Valley suburban sprawl to do his errands.[14] There were breathless predictions about fleets of such self-driving taxis being available in the near future. In 2015, Chris Urmson, then head of Google's self-driving car project and current CEO of self-driving technology company Aurora, expected his then 11-year-old son wouldn't need to get a driver's license.[15] In 2016, ride-sharing company Lyft promised that most of its rides would be in autonomous cars by 2020.[16] That same year, Elon Musk, the owner of Tesla, Inc., declared, "I really consider autonomous driving a solved problem, I think we are less than two years away from complete autonomy, safer than humans."[17] In 2019, Musk claimed he'd have one million fully autonomous robotaxis on the road in the next year.[18]

But as I'm writing this book at the end of 2024, we're still not there yet. Thirty-five years after CMU's prototype, we're getting incrementally closer, but the scale and cost of these cars remain prohibitive. Multiple companies have tried and failed to build a self-driving service. Uber terminated their self-driving car division after killing a pedestrian during an autonomous test drive. Cruise, funded heavily by GM, was effectively shuttered after dragging a pedestrian 20 feet after a collision. Though there were dozens of companies promising self-driving cars and well over $100 billion of investment,[19] the only one delivering rides is Waymo. Many others, like Cruise, have folded.

Bouncing through various corporate structures over the years, Waymo is what used to be Google's self-driving car division. When Google reinvented itself as a conglomerate Alphabet in 2015, it penciled its self-driving car division under the letter W, calling it Waymo. Waymo was soon spun out as an independent company from Alphabet in 2016, though it remains heavily subsidized by the megacorporation. In 2024, Waymo *does* provide a self-driving car service in some capacity. It delivers about 50,000 paid rides a week, primarily in restricted parts of Phoenix and San Francisco. Despite being shuffled around corporate structure and propped up financially for a decade, Waymo has proven that cars with no drivers sort of work!

I say sort of because they remain fundamentally limited. They can't go on highways yet, but they are lobbying to do so soon. Though they technically don't have drivers in the cars, leaked details suggest that crews of remote humans still monitor every driverless Waymo vehicle on the streets. With its varied engineering and human costs, Waymo still loses over $1,000 per ride on these cars.[20] Progress is undeniable, but we remain a long way from the world promised in 1950s advertisements.

Why is autonomous driving so hard?

Part of the problem is that no one can cleanly articulate exactly why it's hard. This lack of articulation is a salient feature for distinguishing problems where people remain better than machines. We don't know how to quantify "good driver" into a clean statistical outcome. Safety would be the lowest possible bar. Cars shouldn't kill people. But, among other things, we also care about comfort in the passenger seat, speed to a destination, and predictability of the car in an urban environment. No one has figured out how to turn this into a clean specification for our complex, unpredictable roads.

Another popular limitation accepted as conventional wisdom is the abundance of what technologists call "edge cases." These are similarly unarticulated parts of the driving model that people didn't envision when putting together their computer code. They could be a new pothole, an unexpected pedestrian, or a street party. Some people think they can add all the possible edge cases to an ordinary statistical model and optimize them away. This is another example of the persistent mindset that all engineering problems can be solved with bigger datasets and computers. Unfortunately, the complexity and uncertainty in the world still far exceed the capabilities of mathematical optimization.

We can look to other twentieth-century successes in automating transportation to see if there's a better path for cars. The most successful autonomous transport, and the only one that requires almost no human intervention, is the elevator. The elevator has closing doors, predictable stops, and simple routing. And it rarely encounters obstacles in the shaft.

Subways are more or less horizontal elevators. The only reason subways aren't fully autonomous is that someone needs to ensure no people get on the track or get stuck in car doors at stations. Automated subway lines have double doors to eliminate these two potential edge cases.

We have airplanes that can take off and land autonomously in adverse conditions, but we create conditions so these planes don't run into each other. A sophisticated air traffic control system ensures that planes not only don't fly near each other but also don't fly into complex weather conditions that would spoil the models used for automation. The system is engineered to minimize uncertainty. The great successes in transportation automation were engineered to be predictable.

Will self-driving cars fully become ubiquitous?[21] This could probably happen tomorrow if enough of the complicated edge cases were engineered out of the system. There would be no edge cases if the cars had dedicated roads without pedestrians or animals. A highway filled only with self-driving cars would also eliminate the need for defensive driving, as the cars could be programmed to work together. It would be faster for everyone if the self-driving cars were platooned together, driving at close distances to each other to maximize fuel efficiency. In other words, we'd already have self-driving cars if we were willing to settle for trains.

Regardless of whether technologists successfully automate the work of taxi drivers, it's not hard to imagine plenty of professions we still have no idea how to automate. Even with tremendous innovations in aviation safety, we still put pilots in the cockpit. Ultimately, we believe pilots can react to situations that are not part of the plan. No matter how well we engineer a system, there are still surprises. Though we might believe otherwise, people are still better at reacting to surprises (if you will, edge cases) than machines automated with mathematical rationality.

Human decision making remains indispensable. We still lean on expert surgeons to perform heart surgery and expert firefighters to fight fires. In these high-stakes situations, decision makers have to act quickly with highly variable uncertainty that doesn't necessarily perfectly match their past experiences. We know some humans perform incomprehensibly well in these situations. The NDM school of thought tries to get a handle on how exactly such proficient experts do such amazing things.

In the 1940s, chess grandmaster and psychologist Adriaan de Groot performed extensive experiments to understand how *people* played chess.[22] Masters were able to identify favorable potential moves far faster than amateurs. This identification allowed them to quickly mentally simulate a few potential future outcomes. And they responded deftly to the dynamics of a game as their opponent adjusted to their strategy. Herbert Simon and his collaborators Allen Newell and Cliff Shaw tried to write computer code to capture these nuances of human decision making in chess. They found their code was far more complex than algorithms derived from Shannon's tree search approaches.[23] They could not include all of the varied strategies and heuristics used by grandmasters. In this way, it seemed that people were *more* sophisticated than computers, and it was harder to identify exactly what procedures they were using to act. Newell, Shaw, and Simon wrongly predicted that future chess solvers would need to deviate from Shannon's program. But they did learn a lot about how humans mastered chess. After decades of play, masters would assemble an internal catalog of tens of thousands of patterns and would be able to solve complex puzzles through mental simulation.[24] De Groot's work thus inspired an investigation into understanding the ways humans were *more sophisticated* decision makers than computers.

Gary Klein dedicated his career to this problem. He spent years observing and interviewing people who make decisions in high-stakes situations under extreme pressure. Doctors, nurses, military commanders at high and low ranks, and athletes and chess players. In all of these cases, Klein and his collaborators found that the pillars of mathematical rationality played no role.[25] Decision makers weren't carefully tabulating statistics, building mathematical models, and solving mathematical optimization problems. Instead, the best human decision makers were like de Groot's chess masters: incomprehensibly sophisticated pattern recognition engines. The more expertise they gathered, the more decisions could be based upon mental manipulation of past experience with intuition, simulation, and storytelling.

Klein's most famous work is his study of firefighters. He embedded himself in a fire department and rode along to many emergency calls. He'd watch firefighters save the lives of people in near-death situations from accidents or decide to evacuate buildings moments before they'd collapse. How did they do this?

Klein and his colleagues observed that firefighters were not optimizing. Rather than sorting through a set of possible actions, they would just evaluate one option and see if it would work. The firefighters insisted to Klein that they never made decisions. When they engaged with an enflamed building, they knew where to enter, where to spray water, and how to find people. They just knew what to do.

Klein saw this sort of reactive intuition in healthcare professionals, pilots, and chess players. From his diverse body of interviews, he formalized a general theory of decision making he called the *recognition-primed decision* (RPD) model. Klein diagrams flow charts about how such recognition-primed decision making operates, but he cannot give us *an algorithm*. It is, at best, only metaphorically algorithmic. There are steps and procedures, but inside those procedures is a complex mix of intuition, mental simulation, metaphor, and storytelling.

They didn't make a list of options and evaluate their relative merits. In case study after case study, Klein saw that the rational decision model couldn't be further from what expert decision makers actually do in high-stress, high-stakes situations.

In situations that seem typical to the decision maker based on their experiences, they mentally simulate the feasibility of actions, acting as soon as they find an action that appears plausibly successful. They proceed if their simulation matches the decision maker's expectations. But if there is an anomaly, they try to explain it through internal storytelling, shifting the recognition to another context.

Both computers *and* people are proficient pattern recognizers. But they recognize patterns in very different ways. Computers match patterns based on statistical regularities. Computers memorize large databases of past experiences, use statistics to interpolate between those experiences, and make decisions based on the past experiences that were most similar. Just as we saw in chapter 5 and Meehl saw in his collection of evidence, this form of computational pattern recognition is powerful. It is the backbone of what we call "artificial intelligence systems" that recommend what to watch on streaming services or function as reliable chat agents. It is also the backbone of statistical prediction that Meehl's lineage has demonstrated superior to expert judgment.

Yet human pattern recognition remains unreplaceable. Human pattern recognition can exploit *difference*. Humans manipulate patterns in their heads to simulate and find anomalies. They can tell stories and warp different patterns so quickly that they can act without thinking. Humans certainly have predictable irrationality, but when well trained and dedicated, they do exceptional things that, as Newell, Shaw, and Simon observed, can't be written down in computer code.

Reconciling the disconnect between the HB and NDM crowds is tricky. It feels like a doctor can assess more than what is fed into the computer or that a therapist can see subtle cues that are valuable for prediction. We think there are many examples of edge cases that statistics can't catch. If these feelings are right, how can we make sense of the HB data suggesting we don't make rational assessments?

First, it should be noted that the HB experiments often artificially placed test subjects into competitions where machines had home field advantage. Behavioral economics studies are all carefully crafted in laboratory settings to elicit the sorts of responses the researchers wanted to see. Consider the experiment I described about the availability heuristic. Most people could only conjure words that start with R, and hence falsely believed more words started with R than had R as the third letter. Lola Lopes notes that out of all 20 consonants, 12 are most frequently found in the first position. If Kahneman and Tversky had used the letter B, people would have been correct in declaring it was more common in the first position. However, if they had provided the letter B, both the availability heuristic and the actual statistical count would have provided the correct answer. For their experiment to distinguish between heuristics and statistics, Kahneman and Tversky had to artificially constrain themselves to the minority of conditions where the heuristic failed.[26] Lopes found several other HB experimental constructions like this and argued that behavioral economics experiments can only draw conclusions if they set up clever but very artificial scenarios. Trapping people in the narrow models we'd associate with computers, people will perform poorly. But these narrow models are not the world where people actually have to act and decide things. So the generalizability of predictable irrationality tends to be very slim. Yes, we're not mathematically rational. No, that doesn't mean people are stupid.

Yet, beyond the HB research, there remains a long train of work inspired by Meehl demonstrating that clinical judgment is worse than statistical judgment on average. The demonstrations considered in these studies have occurred with expert clinicians in real-world settings. The key to the entire clinical-statistical puzzle is those last two words of the first sentence of this paragraph: "*On average.*" The trick that Meehl plays—and that all bureaucrats play—is in the quantification of "better." By better we of course mean better *on average*. Once we decide that things will be evaluated by averages, the game is up. Statistics will win. After all, what is an average other than the most fundamental statistic? If we believe that prediction is possible, and we decide to evaluate based on the percentage of correct predictions, then the best way forward is a method that minimizes the rate of incorrect predictions. Under mild assumptions, this is the best rule you could use to make predictions about the future. That is, since you will be evaluated based on averages, this algorithmic approach is effectively the optimal thing to do.

Meehl summarizes the situation in the last paragraph of his 1954 book. If we subscribe to the bureaucratic utilitarian mindset, the algorithm always wins. No matter how good human intuition might be, it is evaluated against a metric that is a statistical count. Once someone decides that the metric is what needs to be met and that metric needs to be maximized on average, then the best decision *is necessarily statistical.*

> Always, we might as well face it, the shadow of the statistician hovers in the background; always the actuary will have the final word.

In the cases where we want bureaucratic accountability of our predictions, we can lean on an actuarial android.

Where do we draw the line between where we can formulate an algorithm and where we can't? To someone who is not a die-hard rationalist, it is perfectly reasonable to say that describing what humans do is ineffable. Most expertise can't be tidily summarized as numbers, and this is precisely the sort of expertise that remains hard to mechanize. And yet, we know when a human decision is good when we see it. We can find agreement about what we value in human judgment and what makes some humans better than others in occupying different roles.

Perplexingly, an inversion occurs through the articulation of evaluation metrics. If we can measure why humans might be able to outperform machines, then we can build machines to outperform people. On the other hand, if we *can't* cleanly articulate a clean set of actions, outcomes, measurements, and metrics, then we can't mechanize problem solving. It is this digitization, translating the world into the language of the computer, that is needed to automate.

Though their research took them in very different directions, on this point Kahneman and Klein agreed. The two penned a 2009 detente in the *American Psychologist*.[27] For Kahneman and Klein, the real key to when we needed people was when we *couldn't* articulate evaluation. If there isn't a cleanly articulable list of what makes something good, it can't be evaluated.

However, as Meehl noted, once you decide on a metric, the statistician can assess your performance. It's the metric chasing that causes the problems. Kahneman and Klein describe how humans evaluate expertise:

> In most of the situations studied by NDM researchers, the criteria for judging expertise are based on a history of successful outcomes rather than on quantitative performance measures. The most common method for defining expertise in NDM research is to rely on peer judgments. The conditions for defining expertise are the existence of a consensus and evidence that the consensus reflects aspects of successful performance that are objective even if they are not quantified explicitly. If the performance of different professionals can be compared, the best practitioners define the standard.

In other words, they know expertise when they see it. Humans define what expertise is, but it's defined loosely. There is a rough notion of when one person has more mastery than another. However, these comparisons are debatable, and there need not be consensus.

Proponents of mathematical rationality make the mistake of assuming all decision problems are formulable mathematically. Not all uncertainty can be quantified in terms of probability, not all possible outcomes can be accounted for in the clean formulation of an optimization problem, and not all problems are technical problems. While I can't draw the line between what computers can and cannot do, they certainly are not capable of everything. Given that our ideas are so tied to the 1950s conception of computing, there's no suggestion that they will become spontaneously capable of everything at some point in the near future.

Appealing to mathematical rationality as an ideal is odd, as it's not like we couldn't make evidence-informed decisions before the pillars of automated decisions were established. Advances in clean water, antibiotics, and public health brought life expectancy from under 40 in the 1850s to 70 by 1950. From the late 1800s to the early 1900s, we had world-changing scientific breakthroughs in physics, including new theories of thermodynamics, quantum mechanics, and relativity. We built the cars, the airplane, the telephone, the radio, and the movie camera. Technological progress was rapid before the computer age. Even our societal innovations were built without a formal system of rationality. The modern democracy was established, and the modern Constitution was written without optimal decision theory. Wars were fought and won without game theory.

How does the work on human decision-making help us see the limits of automation and understand what makes human decisions special? We might say that automatable knowledge, or *robotic knowledge*, requires concretely measurable outcomes, discrete decisions, and clear rules.[28] Human knowledge is more heuristic, more dependent on local context, less repeatable, and harder to evaluate.

The scope of robotic knowledge is ever-growing. Our impressions of what is and isn't computable have shifted over centuries. Mental arithmetic used to be deemed a sign of exceptional intelligence, but we now don't put much emphasis on it as we all have access to calculators. You can propose that there is no real distinction between human and robotic knowledge. That is, perhaps once we can cleanly articulate all potential moves, potential outcomes, and potential costs, then we can automate all potential decisions. As we've seen so far, it's very hard to build a computer to play chess like a person, but now very easy to build a computer that can play chess.

On the other hand, the self-driving car conundrum demonstrates how some problems can be hard for much longer than we want them to be. The scale required to specify the complexity of all human behavior is still well out of reach. We can only lean on simple models that we can optimize. The assumption of simple, stable prediction underlies our policies. At least for the near term, we're stuck with a split between the human and the robot decision makers.

Practical, inarticulable human knowledge *can* become robotic knowledge. There is a murky boundary that we shift over time, but we are unable to precisely articulate where the boundary is. Of course, this tells us why we'll never be able to see the boundary. If the boundary was known, that might imply that human's capacity to innovate new technological knowledge is also already known and already over. That seems far from the truth. We'll likely continue to innovate, creating new practical and technical knowledge and facilitating new ways to translate between the two forms.

Learning from machines can make us better. Knowing how to use robotic knowledge in our lives can be freeing. The challenge is making sure that we use and deploy it wisely.

7

Cyborg Decision Making

ONE OF the weirder twists in the adoption of a mathematically rational mindset is the assumption that we *ought* to think like computers. The examples I gave in the introduction from academics like Steven Pinker, pundits like Nate Silver, and even sports commentators like Greg Olsen argue rationality is an end in of itself. As I've mentioned at the very start of this book, everyone thinks rationality is good! And if we decide rationality means this very narrow computational definition, we're deciding that any other way to live is probably bad.

Several influential institutions more or less say this directly. Neoclassical economics is premised on the idea that such rationality is an ideal way to organize our society. Utilitarian organizations like Effective Altruism argue that rationality, optimization, and statistical tables can be the basis of what we owe each other. Effective altruists claim you can decide how to give to charity by running a cost-benefit analysis.

No matter how loudly these groups protest, their rational decision making has repeatedly led to bad ends in a real world that consistently defies neat modeling. The financial sector claims a mantle of rationality, but they created the catastrophe that was the subprime mortgage financial crisis. A star of the effective altruism movement, Sam Bankman-Fried, committed massive fraud that he justified because he had decided it was rational to make as much money as possible to give away, even if it meant breaking the law.

More globally, we saw far too much leaning on such rationality when deciding how to deal with the COVID-19 pandemic. In America, there was a public campaign to "Follow the Science," but this resulted in algorithmic policies that, while made to look scientific, were heavily laden with value judgments. In California, we had color-coded scales for reopening schools and businesses based on arbitrary statistics about reported PCR tests. No one was ever told where the various numbers came from, which led to some of the most prolonged school closures in the country.

The problem is that policymaking, whether about decisions for ourselves, our families, our cities, or our countries, is often driven by action bias. Methodological critiques are always met with the canned response, "But we have to do something!" Many will suffer if we don't understand the causes of the opioid crisis, the impact of American abortion bans, the mental health crisis among teenage girls, or how to nonpharmaceutically slow the spread of infectious diseases. We have to do something.

That something is, unfortunately, often just doing more of the same. We fool ourselves by thinking that any meaningful question must be answerable now with the knowledge and technology we have on hand. Technocrats will insist that there is a straightforward path to policy. We just have to define our terms operationally so they are verifiable. We just need to verify things through the proper randomized trial. We know there's a problem of individual differences and sampling errors, but our estimators and stats software will handle these. At this point, we should see the problem. Most important problems don't submit to technocratic systemization. When technocrats insist that any meaningful question must be answerable at this time, they are sadly mistaken.

In this book, we've traced the origins of mathematical rationality from the 1940s to the present. We've explored how its four pillars—optimization, game theory, the randomized clinical trial, and machine learning—arose through a vision of the future in which computers become the "ideal rational agent," enriching human society by computing optimal decisions in the face of uncertainty. But in understanding the technical scope of these tools, we've also seen where and how they fall short. Our modern computational decision making tools, which reduce everything to minimizing statistical risks, can only take us so far. How we act in the face of whatever data we gather will ultimately be up to humans. Even if we have perfectly valid questions we can't answer with cold, statistical empiricism, we still have to make decisions. We still have to do something. So what do we do?

Computers love rules. Their decision systems thrive on clearly articulated goals and constraints, cleanly captured measures of uncertainty, and instructions on how to act given that uncertainty.

By contrast, the most talented human experts thrive on ineffable intuition that often resembles extrasensory perception. Though they can try to mentor trainees in their ways, their best practices are often inarticulable and must be learned through experience. As we described in chapter 6, human expert knowledge is precisely the kind that can't be cleanly written down as a set of rules. Getting the best out of the best people equipped with the best machines thus poses a challenge of interface. The computers want rules and the people thrive on, let's say, improvisation and play.

Unfortunately, as Stigler's optimal diet from chapter 2 highlighted, oversimplifying the world tends to lead to absurd policy prescriptions. You can't mathematically model that food needs some spice or the value of the warming feeling you get when eating grandma's chicken soup. There is no mathematical model of palatability or social value. And herein lies the problem: If you can't quantify it or monetize it, you can't optimize it. Stigler thus mathematizes eating through monetary cost. He, like many economists, leans on money as a metric because it's conveniently numerical. A minimum cost diet is ridiculous, but so are social networks optimized to maximize advertising revenue.

For better or for worse, the trend of the information age has been toward more quantification and to force people to abide by the rules of the machine. Whether through safety checklists, required computerized trainings, or aptitude tests, people are forced through more evaluations, forced to learn more rules, and forced to always deal in numbers.

This path creates considerable friction. In *Streetlights and Shadows*, Gary Klein points out how forcing mechanical rules in high-stakes settings like hospitals or aviation can lead to decision fatigue. Humans dealing with automated rules and systems can get overwhelmed when keeping track of all of the rules. Too many rules can lead to complacency, as a life of following mechanical rules is drudgery. For all these reasons, the adoption of mechanical rules and statistical prediction in high-stakes scenarios must be done with care. Moreover, Klein worries that forcing experts to constantly use algorithmic decision systems can erode their expertise. If a rule is always consulted before a person, the pattern-matching powers that Klein describes get rusty, and they might not be as effective when they are genuinely needed. Algorithms are purportedly introduced so that professionals can focus on the parts of their job where the human impact matters the most, but they have the opposite effect, counterproductively getting in the way of processes that let experts excel.

Additionally, rules are not adaptive to a dynamic and changing world. Statistical rules can become outdated if the statistical regularities on which they are based shift and change over time. Tech companies have to fine-tune their prediction systems daily to adjust to newly generated internet content and ebbs and flows in human interests. Medical risk assessments for predicting illness might stay static for decades, but behavior changes in a medical population, like reduced rates of smoking, can render the old rules invalid. In a dynamic world, we need to dynamically update our automated decision systems. And as of now, only *people* have the ability to notice when the rules go out of date.

There is no better example of what happens when we try to patch old rules with new rules than in bureaucracies. Since the articulation by Franz Kafka, bureaucracy has been viewed as a societal driver of frustration and alienation. But Kafka could only dream of the bureaucracies enabled by large-scale computing machines. The rise of computers coincided with a metastasis of the technocratic state. From their inception, computers have been machines to facilitate bureaucracy. Computers made it easier to count and tabulate statistics. They made it easier to process forms. Communication networks let us send and assemble massive surveys and censuses. Computers were built to count and optimize. And that's what technocratic bureaucracies do, too.

However, in sharp contrast to computers, technocratic bureaucracies are painfully slow. Technocrats blame the sluggishness on the people under their remit. Setting new policies at big workplaces requires time and deliberation and is typically met by resistance from the workforce. Government is even slower, with committees, lobbyists, and interest groups weighing in on complex legislation that, at least in the United States, can only be passed in the first hundred days of a new presidency. We can think of technocratic bureaucracies as massive but slow computers, ones that get reprogrammed every once in a while to set new rules for people.

This metaphor fails because societies are not computer chips. While I noted in chapter 2 that computer chips were often analogized as microscopic cities, chips were always designed to be hermetically sealed and perfectly controlled. This is what made them optimizable. Real societies, on the other hand, have people. While it's convenient to model and view the population, its health, and its market flows as mathematical abstractions, these run into the limits of the messiness that people bring to bear.

This doesn't stop policymakers from trying to get rid of this messiness.

A popular policy approach inspired by behavioral economics asserts that policies can be made to play to people's biases so that they fall in line. If people are predictably irrational, policies must be made to save them from themselves. This is the mindset of "libertarian paternalists" like Obama advisor Cass Sunstein and Nobel Prize Winner Richard Thaler. Thaler and Sunstein wrote a book, *Nudge*, advocating for policies that would steer people to more mathematically rational decisions, with the idea that forcing people into the model of the ideal rational agent would lead to more prosperity.[1]

"Nudge politics" downplays people's ability to figure things out on their own. It tries to socially engineer systems that politely push people to behave the way policymakers want them to behave. It is based on the idea that these politicians know better than everyone else because they are following the guidelines of ideal rationality. But if mathematical rationality is limited and societies are not computer chips, it doesn't make much sense for us to tolerate a bunch more rules, no matter how subtle those rules are.

I bring up bureaucracy not only because it has grown in the age of the computer. It also gives us one of the best illustrations of the interplay between rules and play. In his book *The Utopia of Rules*, David Graeber tries to understand modern society's passion for bureaucracy through an anthropological lens.[2] Graeber argues bureaucracy functions as a societal game. Graeber gives a more expansive definition of a game that still captures the ideas central to von Neumann and Morgenstern's formulation discussed in chapter 3:

> First, they are clearly bounded in time and space, and thereby framed off from ordinary life. There is a field, a board, a starting pistol, a finish line. Within that time/space, certain people are designated as players. There are also rules, which define precisely what those players can and cannot do. Finally, there is always some clear idea of the stakes, of what the players have to do to win the game. And, critically: that's all there is. Any place, person, action, that falls outside that framework is extraneous; it doesn't matter; it's not part of the game. Another way to put this would be to say that games are pure rule-governed action.

This closely parallels what a game theorist would say, but Graeber departs from von Neumann and Morgenstern by asking why people play games in the first place. The obvious answer is because they are fun. But what makes them fun? Graber argues that games are satisfying because they are the only part of our existence where rules aren't ambiguous. Instead of social norms about politeness and etiquette, games have explicit rules. And it's stressful for us when we don't understand the rules of some new environment, whether it be starting a new job or getting into a fight with a loved one. There, we have implicit social norms that we can learn over time, but these norms are ambiguous and constantly shifting. It's only in games that the rules are clear. It's also only in games that it's possible to decide what a win is. Remember, being able to declare a win was also central to Meehl's characterization of when we could apply statistics to make decisions.

The other central part of games that makes them fun is that both parties agree to the rules. We enter into games willingly. While von Neumann and Morgenstern's formulation of games focuses on rules, it hides the fact that we *play* games.

For Graeber, play is any free-form creativity. Clearly we use this notion of play when we literally play inside the rules of a game. Games inherently balance rules against play, some allowing for more play than others. In board games, the moves are so restricted that creative play strategy can only emerge over long sequences of moves stitched together. However, in other games, like charades, the only rule is that the player can't speak, and the creativity of expressing an idea through movement is what wins the game.

But play is more general, more broad. Play can create new games. Arguing about rules can be a form of play. Play can just be an end in itself, creativity for the sake of creation. But play can also be dangerous. "Horseplay" can lead to injuries. Playing with matches might cause fires. There is a fine line between joking around with someone and harassing them. Sometimes someone's improvisation gives them advantages that others view as unfair. Play can be frightening and potentially destructive.

So in our social systems, if we want fairness and transparency, we need rules. Graeber argues that this is an unseen force in modern participatory democracies. Leaning on mathematical rationality implies predictability and stability. Uncertainty is quantified and managed by declaring the rules and outcomes and enforcing actions that maximize societal well-being. This seems good! Rules ensure a level playing field. By enforcing "rationality" at a social scale, ones where the ends are liberal values, we end up leaning on rules. Paradoxically, in pursuit of more fair, transparent, predictable outcomes, we lean on more explicitly written and enforced rules. You increase transparency and clarity by removing personal authority. In striving for more fairness, Graeber argues we end up sacrificing our ability to be creative and choose our own paths and destinies. A fear of play leads to an expansion of rules and more bureaucracy.

Graeber's illustrated tension between rules and play parallels the tension between the "heuristics and biases" (HB) and "naturalistic decision making" (NDM) schools we encountered in chapter 6. Gary Klein describes the former as viewing humans as hazards and the latter as viewing humans as heroes. By enforcing "rationality" in health care, we create transparent rules that reduce certain errors and create accountability. However, we also end up with a bureaucratic system that constrains caregivers with checklists. The rules get made, and they then become impossible to change. The path to bureaucratic hell is paved with many good intentions.

Graeber argues that the obsession with mathematical rationality and the computer distracted from more poetic goals of the Cold War Era. The Moon landing was in 1969. We haven't made advances in human spaceflight since then. We don't have teleporters or flying cars. But we have powerful tools of rules and virtualization. These tools enabled the exponential growth of bureaucracy in the postwar period.

Rules, it turns out, are one of our most potent tools for taming uncertainty. If we add enough rules, society becomes more like the utopian computer chip. But rulemaking is a tradeoff between a greater good and a personal good. How can we decide when enough is enough? Computers have made it an ideal to be rational. We can view people as rule-following. We create more rules. This creates comfort and authority. But it is at the cost of creativity and play. As Graeber writes:

> Whether motivated by a faith in "rationality" or a fear of arbitrary power, the end result of this bureaucratized notion of freedom is to move toward the dream of a world where play has been limited entirely—or, at best, boxed away in some remote location far from any serious, consequential human endeavor—while every aspect of life is reduced to some kind of elaborate, rule-bound game. It's not that such a vision lacks appeal. Who hasn't dreamed of a world where everyone knows the rules, everyone plays by the rules, and—even more—where people who play by the rules can actually still win? The problem is that this is just as much a utopian fantasy as a world of absolute free play would be.

The problem with games is that they are fun for a while. It's great to know there are rules. But they can be stifling. We need free play too. Unoptimizable anarchy and refusal to play by the rules are what keeps life worth living. We need the ability to make choices that are fulfilling for the sake of fulfillment. And perhaps this is the fundamental question of how we move forward out of the computer age: How exactly can we balance rules and play? To begin to answer this question, let's ask it in the context of the action-bias conundrum I introduced at the beginning of this chapter. When we have to make an actual decision that impacts real people, we often find mathematical rationality and its attendant rules increase rather than reduce the uncertainty in our decision making.

In a polemic, "How Science Makes Environmental Controversies Worse," political scientist Daniel Sarewitz worries that computational, statistical, and scientific thinking often makes decision making worse by introducing an illusion of authority.[3] Though Sarewitz's title focuses on climate science, he gives examples in agriculture and political science that buttress a broader argument.

Science and evidence are always uncertain. Models are always approximations of reality. Predictions from the past can only say so much about the future. Given the persuasions of any particular scientist, scientific theories can be attacked from a variety of different arguments. Two scientists might disagree about methods, conditions under which scientific results hold, or the various metrics each uses to evaluate a particular problem. Since models are always idealizations of the world and uncertainty quantification is itself highly uncertain, uncertainty will always remain, as long as there's the will of a large enough group of other scientists to keep the arguments going.

Nothing provides such will more than politics. This means that whenever a scientific question becomes political, asking if science says we should do something only increases uncertainty. We mask a problem of *ought* by a problem of *is*. As long as there are camps on either side, we sink into a morass where uncertainty becomes larger and larger. The scientific method almost ensures you can never get a scientific answer to a political question.

Perhaps Sarewitz's most compelling example is the recount of the 2000 election between George W. Bush and Al Gore. The election would be decided by the difference between the votes that the candidates received in the state of Florida. In essence, the question was determining a single integer. The problem was that the vote was so close, a difference of about 900 out of six million votes cast on its first count, that experts had to be brought in to audit whether the count was correct. The more experts that were brought in, the more questions were raised about voter intent, about how to determine if a ballot was punched, and about the appropriate statistical error analysis. The experts' vacillations changed no one's mind. Sick of the expert stalemate, the Supreme Court voted 5 to 4 to elect George Bush through a convoluted argument about the Equal Protection Clause of the Fourteenth Amendment. Their strained legal reasoning isn't important to the story here, but their means are. In the end, the problem was decided by politics, not science, rationality, or computation.

Sarewitz's conclusion is counterintuitive but points to the disutility of uncertainty quantification in policymaking. Experts can see whatever they want to see if they try hard enough. Presentations from two competing camps only result in more questions, more heat, and little clarification.

So what do we do? Sarewitz concludes that decisions under uncertainty are thus necessarily more about ethics and values than optimization and uncertainty quantification. We decide based on what we believe ought to be true. We decide based on what we'd like to make real. We decide based on the sort of society we want to be. Statistical measurement can't tell us any of these things. And it never will be able to.

We have to resist the urge to make value judgments look scientific by adding numbers and charts. Such quantification, even if it is invalid, makes subsequent decisions look more rational. This is dangerous, and we must push back against it whenever we see it. We can use a language of rationality and science to claim authority on how best to decide things, but at some point we need to come to terms with the fact that what we ought to do—what our values say we should do—isn't going to neatly line up with an actuarial table.

Sarewitz's argument entreats us to understand the limits of technocratic knowledge. Neil Postman argued in the 1980s that we were becoming too reliant on siloed expert knowledge and had convinced ourselves it could play an impossible role:

> The role of the expert is to concentrate on one field of knowledge, sift through all that is available, eliminate that which has no bearing on a problem, and use what is left to assist in solving a problem. This process works fairly well in situations where only a technical solution is required and there is no conflict with human purposes—for example, in space rocketry or the construction of a sewer system.... [I]t is disastrous when applied to situations that cannot be solved by technical means and where efficiency is usually irrelevant, such as in education, law, family life, and problems of personal maladjustment.[4]

Postman articulated exactly what Sarewitz concluded a decade later. When we face scientifically unanswerable questions, we need to be open about making decisions based on our values, not on our science and technology.

Even when faced with disagreement and irreducible uncertainty, at some point we must choose. The only decision makers that can choose are those with volition. And those are people, not machines. Machines can make decisions. Only people can choose.

This distinction between decisions and choices was made by Joseph Weizenbaum in his 1976 book *Computer Power and Human Reason*.[5] Weizenbaum had been engaged in artificial intelligence research since the early 1960s, though he was always more interested in *the appearance* of intelligence than in intelligence itself. In 1962, following the lead of other AI researchers, he dipped his toe into gaming, designing a computer program to play a baby version of Go, "Gomuku," that could beat novices. Rather than crowing about how machines would become superintelligent, however, Weizenbaum's work on Gomuku is entitled "How to Make a Computer Appear Intelligent."[6]

But he's most famous for designing the first famous chatbot, ELIZA, named after the character in George Bernard Shaw's *Pygmalion*. In the play, Eliza is a poor girl who works in the flower shop and is the subject of a cruel bet that she could pass for a duchess if taught proper speech. Demoed in 1966, ELIZA was the first conversational chatbot, which Weizenbaum hypothesized could also be taught to speak (or at least type) in convincing proper language.[7]

Weizenbaum built ELIZA to parody a psychotherapist. It engaged in active listening, mostly repeating back what the human put into the machine. Here's a sample exchange provided by Weizenbaum:

> Men are all alike.
> **IN WHAT WAY**
> They're always bugging us about something or other.
> **CAN YOU THINK OF A SPECIFIC EXAMPLE**
> Well, my boyfriend made me come here.
> **YOUR BOYFRIEND MADE YOU COME HERE**
> He says I'm depressed much of the time.
> **I AM SORRY TO HEAR YOU ARE DEPRESSED**

It would go on like this. Weizenbaum thought ELIZA was a neat demonstration of the power of computing, easily accessible because everyone understood how to have a conversation. Nothing could prepare him for people's over-the-top reactions.

Many computer scientists and psychologists thought Weizenbaum had cracked the key to understanding language with his simple, rule-based system. Published journal articles exclaimed that computers would replace the psychologist. Weizenbaum was aghast at the "mechanical conception of man." Therapists weren't supposed to be algorithms. They were healers engaged in human interaction with their patients. In laypeople, Weizenbaum observed emotional attachments to ELIZA. People would anthropomorphize the software program in an attempt to heal themselves. Weizenbaum considered these reactions delusional.

You could replace ELIZA with 2023's ChatGPT and make all of the same arguments. In fact, people did make exactly these arguments, often citing ELIZA as a cautionary tale.

Weizenbaum spent a decade thinking about how to best explain why ELIZA wasn't intelligent and why computers could never replace people. He concluded that the fundamental confusion was a conflation between decisions and choices. Weizenbaum cites Michael Polanyi, who, in arguing against Soviet social planning, asserted that "a mechanical conception of man and history ... denied any grounds for claiming freedom of thought."[8] For Weizenbaum, freedom of thought was indeed the key distinction. Unlike machines, humans had freedom of thought, a moral calling, and an ability to choose. Machines can decide, but only humans can *choose*.

Weizenbaum's book concludes with an argument not dissimilar to the one you've read here. His final chapter is "Against the Imperialism of Instrumental Reason." Against Mathematical Rationality, if you will. Weizenbaum argues that by harnessing mathematical rationality into a governance system, people become nothing more than cogs in a machine. They feel like they themselves become powerless robots. While proponents of rationality point to the tremendous postwar progress, Weizenbaum asks if the price is that humans must serve rationality, rather than the other way around.

He rejects this ideology and issues a call to arms that is still heard today, though often bulldozed by technological progress. Weizenbaum argues against technological inevitability. "We can count, but we are rapidly forgetting how to say what is worth counting and why." He wants people to embrace their power to choose. There are some projects, for instance, that shouldn't be done. He argues against building speech recognition systems because of their inevitable integration into systems of surveillance. He certainly wasn't wrong about this. Whether mass surveillance is good or evil remains a heated political argument today, with different communities engaging in political wars about how our mobile devices track our every move. Weizenbaum argues that the political self, the ethical self, and the courageous self cannot be separated from our existence and can't be mathematized. We have to embrace our values and choose what ought to happen. We can't look up the solution to political problems in an actuarial table. "What could it mean to speak of risk, courage, trust, endurance, and overcoming when one speaks of machines?"

Most books on technology either take the side that all technology is bad or all technology is good. This isn't one of those books. Such books focus too much on harm and not enough on limits. Limits are more empowering. Throughout the book, I've maintained that mathematical rationality is limited in what kinds of problems it is best placed to solve but has sweet spots that have yielded remarkable technological advances.

The role of this technology in value-laden decision making remains complex and fraught with danger. Some problems are not amenable to statistical tabulation and cost-benefit calculation. Perhaps the best takeaway from this book is understanding how mathematical rationality and all its prescriptions are artificial and made by people. Since these rules are all artificial, there can't be a single "right" way to make decisions and evaluate interventions. All algorithmic rulemaking is necessarily value-laden and flawed.

However, I believe in participatory decision making. A key feature of rules is that they are *changeable*. Rules can be adapted so that processes better align with what we value. One redeeming aspect of bureaucracy is its potential to get stakeholders to agree to the game by articulating the rules cleanly. Calling statistical methods "regulations" centers our agency in using and changing them. I have argued that we can't compute our way to utopia. But we can be honest about what we are doing as computational scientists, engineers, and policymakers. We can be frank about how our work impacts people.

If you feel like one of Weizenbaum's metaphorical cogs in the machine, disenfranchised from choosing, collective action remains a formidable solution. The power of human choice is only amplified by working together, whether it be through organizing to change your work experience or how your government rules. It's not easy. It's not straightforward. But the best way to push back against machine rationality is to choose to gather together our collective humanity on the small scale.

AFTERWORD

WRITING THIS book has been a life-changing, enriching process, and I hope you enjoyed where I ended up. As part of an extended ten-year midlife/midcareer crisis, I have been determined to see through the transition from writing information science to writing *about* information science. Despite scouring through ample academic advice, I didn't find an instruction manual for how to proceed.

Whereas my previous books emerged from the classroom, I wasn't sure from the outset how to get after my questions about the history and philosophy of automated decisions in a course. I had to experiment with method as I went along. My process ultimately became highly interactive and social, which comes as no surprise to those who know me. As a result, there is a very long list of amazing people who specifically shaped the narrative here and sequentially steered me in the right directions. I am deeply indebted to all of them.

It was amazing to talk to some of the people centrally involved in making this history. I'd like to thank Alberto Sangiovanni-Vincentelli for recounting the revolution in electronic design automation. Manfred Morari shared helpful perspectives on the end of the height of the automatic control era of the 1980s. I'd also like to thank Daphne Koller and Michael Bowling for sharing their recollections of and perspectives on their incredible breakthroughs in game theory. I received fascinating insights from individuals who have applied randomized trials in industrial settings, including Matt Folz, Ryan Giordano, Omkar Muralidharan, and Chris Wiggins. I'm grateful to John Platt and Csaba Szepesvári for our email conversations about the state of machine learning in the 1980s and the role of competitive testing in our field.

I also received many valuable pointers from fellow fans of the intellectual history of mathematical rationality. During the pandemic, Moritz Hardt shared my passion for digging up the early developments in the history of machine learning. Putting together the pieces of the story of Bill Highleyman and his data, which first appeared in the Datasets chapter of our book, *Patterns, Predictions, and Actions*, was a joy. Murat Arcak pointed me to several interesting developments in early optimal control. Gabriele Farina and Gergely Neu tutored me on the modern theoretical tools for computational game play. Ziad Obermeyer and Adam Cifu imparted their perspectives on the role of randomized evidence in medicine. Sasha Rakhlin and Maxim Raginsky helped me understand some of the parallel developments in machine learning in the 1960s Soviet Union.

Scott Shenker was a tirelessly supportive sounding board for the overall narrative structure of the book and helped me determine which parts to keep and which to save for later. I had several helpful guiding conversations with Francesco Borrelli, Jordan Ellenberg, Eric Jonas, Jay Caspian Kang, Ali Rahimi, Lawrence Recht, and Christopher Ré. I'd also like to thank George Dyson for his insights on early computer development and the impact of the peculiar personalities of Shannon, Wiener, and von Neumann. The Miller Institute for Basic Research in Science at UC Berkeley generously sponsored me for a research fellowship at the beginning of my writing process, and my discussions about this book and related research with their community of faculty and fellows were invaluable.

To synthesize all of this gathered information, I approached this text as a working paper, iterating on much of it in public. I used speaking engagements as fake deadlines to turn some of the nascent ideas into hour-long talks. I also used my blog, argmin.net, as a sketch pad, where I first workshopped much of the text that you read here. Talks and blogs are ephemeral, but interactions with the audience helped to shape and hone my arguments. All of those early interlocutors, in person and online, helped shape this final archival form. I'm thankful to all of you.

I'd like to thank Shamik Dasgupta for envisioning and coteaching a seminar on the topics of this book in spring 2024 at UC Berkeley. In the end, this book didn't emerge out of the classroom, but it was finished there. The course also helped ground the intellectual history of automation and decision making in broader philosophical thinking. A big thank you to all the seminar students for their feedback and thoughts on the draft chapters we read.

I'm grateful to Siva Balakrishnan, Damek Davis, Daniel Hsu, and Katy Lankester, who all read drafts of this book and provided tremendously helpful feedback. I'd also like to thank my research group, Mihaela Curmei, Jessica Dai, Sara Fridovich-Keil, Paula Gradu, Chris Harshaw, Holly Jackson, and Deb Raji. My group not only read the book, but also sat through many early versions of practice talks and had to listen to me blather on about related themes for the past several years. A special thank you to Jess, whose editorial eye made the argumentation much crisper.

Thanks to Hallie Stebbins, my editor at Princeton, for helping me shape this book into a readable, coherent story and guiding me through the editorial process.

And let me end with the most important person, Lauren Kroiz, whom I am beyond lucky to be married to. Lauren, a professor of art history at UC Berkeley, helped me with honing historical methodology and best archival research practices. She also heard many of the half-baked ideas of early chapters that didn't fit together, and helped me piece together more coherent arguments. This book wouldn't have been possible without such a generous partner and mentor. I can't thank her enough. She's the best.

NOTES

Chapter 1

1. Pinker, Steven. *Rationality: What It Is, Why It Seems Scarce, Why It Matters.* Penguin Publishing Group, 2021.

2. That he needed to throw the word "true" in there already shows how hard it is to pin down a definition of rationality. Indeed, Pinker spends the next several pages arguing about what might be true or not.

3. Silver, Nate. *On the Edge: The Art of Risking Everything.* Penguin Publishing Group, 2024.

4. The Pat McAfee Show, January 23, 2024. https://www.youtube.com/live/lAqJ4CBq2G4?t=2886.

5. Baldwin, Roger R., Wilbert E. Cantey, Herbert Maisel, and James P. McDermott. "The Optimum Strategy in Blackjack." *Journal of the American Statistical Association* 51, no. 275 (September 1956): 429–39. https://doi.org/10.1080/01621459.1956.10501334.

6. Postman, Neil. *Technopoly: The Surrender of Culture to Technology.* Penguin Random House, 1992.

7. Bethlehem, Jelke. "The Rise of Survey Sampling." Discussion Paper 09015, ISSN 1572–0314. CBS/Statistics Netherlands, 2009.

8. Edwards, Paul N. *The Closed World: Computers and the Politics of Discourse in Cold War America.* MIT Press, 1997.

Chapter 2

1. Army Air Forces Statistical Digest, World War II. https://apps.dtic.mil/sti/citations/ADA542518.

2. Hay, David Lowell. "Bomber Businessmen: The Army Air Forces and the Rise of Statistical Control, 1940–1945." Ph.D. dissertation, University of Notre Dame, 1994. https://www.proquest.com/docview/304115991/abstract/1D50E7C5C98E4299PQ/1.

3. Dantzig, George B. "The Need for High Speed Electronic Computers for Programming." Presented to the RBD Committee of Basic Physical Sciences, August 22, 1949.

4. Dantzig, George B. "Reminiscences about the Origins of Linear Programming." *Operations Research Letters* 1, no. 2 (April 1982): 43–48. https://doi.org/10.1016/0167-6377(82)90043-8.

5. Dantzig, George B. "Programming in a Linear Structure." Report of the September 9, 1947 meeting in Madison, *Econometrica* 17 (1949): 73–74.

6. Dantzig, George B. "Maximization of a Linear Function of Variables Subject to Linear Inequalities." In *Activity Analysis of Production and Allocation*, edited by T. C. Koopmans. John Wiley & Sons, New York, 1951. 339–47.

7. "The term 'simplex' technique arose in a geometric version of this development which assumes one of the m equations [of the linear constraints is that the variables in the policy sum to 1]. [With the extra constraint] a basis may be considered one of the faces of a simplex." Dantzig notes that you can reformulate any problem to have this constraint. But this is unnecessary and not required in his algorithm. Basis is a more evocative colloquial term, and also easier to explain to people who are not mathematicians.

8. Stigler, George J. "The Cost of Subsistence." *Journal of Farm Economics* 27, no. 2 (1945): 303–14. https://onlinelibrary.wiley.com/doi/10.2307/1231810.

9. Rowena S. Carpenter and Hazel K. Stiebeling. "Diets to Fit the Family Income." *USDA Farmers' Bulletin* 23 (1936): ii + 38 pp.

10. There is a slight technical inaccuracy in Dantzig's solution. Stigler assumed that when you bought goods in bulk, you were not allowed to purchase fractional amounts of goods. You had to buy flour by the pound and half pounds weren't allowed. You had to buy beans by the can. This means that the linear program must be solved with the additional constraint that the solution have integer coordinates. This is called an *integer program*. Using integer programming, the minimum cost is $39.74, meaning Stigler was only off by 19 cents.

11. There is a Georgian bean pie called *lobiano* that is quite delicious. It contains a cup of flour for every cup of navy beans, but also features a tablespoon of lard or bacon drippings. A good recipe can be found in: Alford, Jeffrey and Naomi Duguid. *Home Baking: The Artful Mix of Flour and Traditions from Around the World*. Artisan, 2003. 93.

12. "Tax-supported bureaucrats and professors may also have another reason for certain of their practices."

13. Dantzig, "Reminiscences about the Origins of Linear Programming."

14. Dantzig, "Need for High Speed Electronic Computers."

15. Northrop Aircraft Corporation's BINAC machine was a stored-program computer released a year earlier in 1949. The BINAC was designed by the same team as

the seminal ENIAC computer from the University of Pennsylvania. But the BINAC computers never shipped in functional states; the SEAC was the first stored-memory computer that could actually run computations.

16. Votaw, D. F., and A. Orden. "The Personnel Assignment Problem." Air Force PROJECT SCOOP MANUAL no. 10 (April 1952): 159–63.

17. Orchard-Hays, William. "Evolution of Linear Programming Computing Techniques." *Management Science* 4, no. 2 (January 1958): 183–90. https://doi.org/10.1287/mnsc.4.2.183.

18. Kalman, R. E., and R. W. Koepcke. "The Role of Digital Computers in the Dynamic Optimization of Chemical Reactions." In *Papers Presented at the March 3–5, 1959, Western Joint Computer Conference*, 107–16. IRE-AIEE-ACM '59 (Western). New York, NY, USA: Association for Computing Machinery, 1959. https://doi.org/10.1145/1457838.1457857.

19. Kalman and Koepcke wrote this objective as minimizing the mean-squared error between the state of the output stream and the desired steady-state concentrations.

20. As evidence that these earliest programming languages captured some of the most important abstractions, most mathematical and scientific computing is still based on code written in FORTRAN.

21. Dantzig, George B. "Linear Programming Under Uncertainty." *Management Science* 1, no. 3/4 (April–July 1955): 197–206.

22. Frank, Marguerite, and Philip Wolfe. "An Algorithm for Quadratic Programming." *Naval Research Logistics Quarterly* 3, no. 1/2 (1956): 95–110. Markowitz, Harry. "The Optimization of a Quadratic Function Subject to Linear Constraints." *Naval Research Logistics Quarterly* 3, no. 1/2 (1956): 111–33.

23. Dantzig, George B. "On the Significance of Solving Linear Programming Problems with Some Integer Variables." *Econometrica* 28, no. 1 (January 1960): 30–44. https://doi.org/10.2307/1905292.

24. Rosen, J. B. "The Gradient Projection Method for Nonlinear Programming, Part I: Linear Constraints." *Journal of the Society for Industrial and Applied Mathematics* 8, no. 1 (March 1960): 181–217. https://doi.org/10.1137/0108011.

25. Fletcher, R., and C. M. Reeves. "Function Minimization by Conjugate Gradients." *The Computer Journal* 7, no. 2 (January 1964): 149–54. https://doi.org/10.1093/comjnl/7.2.149.

26. Polyak, B. T. "Some Methods of Speeding Up the Convergence of Iteration Methods." *USSR Computational Mathematics and Mathematical Physics* 4, no. 5 (1964): 1–17.

27. Broyden, C. G. "The Convergence of a Class of Double-Rank Minimization Algorithms, 1: General Considerations." *Journal of the Institute of Mathematics and Its*

Applications 6 (1970): 76–90. https://doi.org/10.1093/imamat/6.1.76; Fletcher, R. "A New Approach to Variable Metric Algorithms." *Computer Journal* 13, no. 3 (1970): 317–322. https://doi.org/10.1093/comjnl/13.3.317; Goldfarb, D. "A Family of Variable Metric Updates Derived by Variational Means." *Mathematics of Computation* 24, no. 109 (1970): 23–26. https://doi.org/10.1090/S0025-5718-1970-0258249-6; Shanno, David F. "Conditioning of Quasi-Newton Methods for Function Minimization." *Mathematics of Computation* 24, no. 111 (1970): 647–56. https://doi.org/10.1090/S0025-5718-1970-0274029-X.

28. Nelder, J. A., and R. Mead. "A Simplex Method for Function Minimization." *The Computer Journal* 7, no. 4 (January 1965): 308–13. https://doi.org/10.1093/comjnl/7.4.308.

29. Fletcher, R. "A New Approach to Variable Metric Algorithms." *The Computer Journal* 13, no. 3 (January 1970): 317–22. https://doi.org/10.1093/comjnl/13.3.317.

30. Bryson, A. E., and Stanley E. Ross. "Optimum Rocket Trajectories with Aerodynamic Drag." *Journal of Jet Propulsion* 28, no. 7 (July 1958): 465–69. https://doi.org/10.2514/8.7355.

31. Bryson, A. E., and W. F. Denham. "A Steepest-Ascent Method for Solving Optimum Programming Problems." *Journal of Applied Mechanics* 29, no. 2 (June 1962): 247–57. https://doi.org/10.1115/1.3640537.

32. Moore, Gordon E. "Cramming More Components onto Integrated Circuits." *Electronics* 38, no. 8 (April 19, 1965): 114–17.

33. Sangiovanni-Vincentelli, Alberto. "My 50-Year Journey from Punched Cards to Swarm Systems." In *Proceedings of the 2019 International Symposium on Physical Design (ISPD)*, 123–25. San Francisco: ACM, 2019. https://doi.org/10.1145/3299902.3311071. See also: Sangiovanni-Vincentelli, A. "Corsi e Ricorsi: The EDA Story." *IEEE Solid-State Circuits Magazine* 2, no. 3 (2010): 6–25. https://doi.org/10.1109/MSSC.2010.937693.

34. Stein, G. "Respect the Unstable." Bode Lecture at the IEEE Conference on Decision and Control. 1989. Modified transcript reprinted as: Stein, Gunter. "Respect the Unstable." *IEEE Control Systems Magazine* 23, no. 4 (August 2003): 12–25. https://doi.org/10.1109/MCS.2003.1213600.

35. It really is hard to exaggerate Shannon's influence on modern technology! Shannon will play a pivotal role in chapters 3 and 5. But I'm barely going to do justice to all of the important work he did.

36. Artificial intelligence fabulist Marvin Minsky once berated me, saying my work in machine learning was nothing more than 1930s optimization research. He was right. I was working on optimization. I'm sure he's rolling over in his grave now that it turns out that all of AI is nothing but 1930s optimization research.

Chapter 3

1. Von Neumann, John, and Oskar Morgenstern. *Theory of Games and Economic Behavior*. Princeton University Press, 1944.

2. Von Neumann, J. "Zur Theorie der Gesellschaftsspiele." *Mathematische Annalen* 100 (1928): 295–320.

3. Von Neumann of course did not have the luxury of looking at what happened in physics from 1944 to 2023, in which case he might have had a bit more humility.

4. More discussion on von Neumann's complicated relationship with randomness can be found in: Mirowski, Philip. "What Were von Neumann and Morgenstern Trying to Accomplish?" In *Toward a History of Game Theory*, Annual Supplement to volume 24, *History of Political Economy*, edited by E. Roy Weintraub. Duke University Press, 1992.

5. Bellman, Richard, and David Blackwell. "Some Two-Person Games Involving Bluffing." *Proceedings of the National Academy of Sciences* 35, no. 10 (October 1949): 600–605. https://doi.org/10.1073/pnas.35.10.600.

6. Vopson, Melvin M. "The Information Catastrophe." *AIP Advances* 10, no. 8 (August 2020): 085014. https://doi.org/10.1063/5.0019941.

7. Shannon, Claude E. "XXII. Programming a Computer for Playing Chess." *The London, Edinburgh, and Dublin Philosophical Magazine and Journal of Science* 41, no. 314 (March 1950): 256–75. https://doi.org/10.1080/14786445008521796.

8. Bernstein, Alex, and Michael de V. Roberts. "Computer v. Chess-Player." *Scientific American* 198, no. 6 (June 1958): 96–107.

9. Weiss, E. A. "Biographies: Eloge: Arthur Lee Samuel (1901–90)." *IEEE Annals of the History of Computing* 14, no. 3 (1992): 55–69. https://doi.org/10.1109/85.150082.

10. Samuel, A. L. "Some Studies in Machine Learning Using the Game of Checkers." *IBM Journal of Research and Development* 44, no. 1.2 (January 2000): 206–26. https://doi.org/10.1147/rd.441.0206.

11. Sutton and Barto discuss why Samuel's algorithm wasn't a valid implementation of temporal differencing. Because of the incorrectness of his algorithm, Samuel had to rely on heuristics to force Alpha to converge upon a strategy. Sutton, Richard S., and Andrew G. Barto. *Reinforcement Learning: An Introduction*. MIT Press, 2018.

12. The game was discussed by IBM on its online archive, which has unfortunately been removed from the IBM website. Here is a link to an archived webpage: https://web.archive.org/web/20230504101300/https://www.ibm.com/ibm/history/ibm100/us/en/icons/ibm700series/impacts/.

13. A lot of salty commentary can be found at the Chinook project. To be fair, the Chinook team actually solved checkers 30 years later. http://webdocs.cs.ualberta.ca/~chinook/project/legacy.html.

14. There was the admirable goal of helping the national interest. But military funding was a core part of mathematics research in the early postwar period. In many regards, RAND wasn't special. Think tanks staffed by mathematicians and funded by the army were commonplace, including the Army Math Research Center at Madison, famously the target of a bombing by antiwar activists in 1970.

15. Bellman, Richard, and David Blackwell. "A Bomber-Fighter Duel (I)." RAND Corporation Research Memorandum RM 165, 1949.

16. Leonard, Robert. *Von Neumann, Morgenstern, and the Creation of Game Theory: From Chess to Social Science, 1900–1960.* Cambridge University Press, 2010. 314.

17. Radzik, Tadeusz. "Results and Problems in Games of Timing." *Statistics, Probability, and Game Theory*, IMS Lecture Notes—Monograph Series 30 (1996): 269–92.

18. https://wrpsa.com/rock-paper-scissors-tournaments/.

19. Look, I can't tell if this is real or not, but I admire it nonetheless. https://www.youtube.com/watch?v=XAY7rKQfVtM.

20. Leonard, *Von Neumann, Morgenstern*, 320–21.

21. Von Neumann and Morgenstern, *Theory of Games*, 2nd ed. (1947), 87, n3.

22. For an extended discussion of the many unpredictable outcomes of Ultimatum, see Gale M. Lucas et al., "Against Game Theory." In *Emerging Trends in the Social and Behavioral Sciences: An Interdisciplinary, Searchable, and Linkable Resource*, edited by Robert A. Scott and Stephen M. Kosslyn, 1–16. Wiley, 2015. Available at https://scholarship.law.duke.edu/faculty_scholarship/3479.

23. Leonard, 315.

24. https://www.computerhistory.org/chess/orl-4345632d88ad1/.

25. Weber, Bruce. "Swift and Slashing, Computer Topples Kasparov." *The New York Times*, May 12, 1997, 1.

26. Schaeffer, Jonathan, Neil Burch, Yngvi Björnsson, et al. "Checkers Is Solved." *Science* 317, no. 5844 (September 14, 2007): 1518–22. https://doi.org/10.1126/science.1144079.

27. Berliner, Hans J. "BKG—A Program that Plays Backgammon." Technical Report of the Carnegie-Mellon University Computer Science Department, 1977; Berliner, Hans J. "Backgammon Computer Program Beats World Champion." *Artificial Intelligence* 14, no. 2 (September 1980): 205–20. https://doi.org/10.1016/0004-3702(80)90041-7.

28. By the mid-1990s, it was evident that TD learning was an approximation method to solve Shannon's general dynamic programming formulation of gameplay, and this helped further innovations in the field. See the discussion in Bertsekas, Dimitri P., and John N. Tsitsiklis. *Neurodynamic Programming*. Athena Scientific, 1996.

29. Kocsis, Levente, and Csaba Szepesvári. "Bandit Based Monte-Carlo Planning." In *Machine Learning: ECML 2006*. https://doi.org/10.1007/11871842_29.

30. Gelly, Sylvain, Levente Kocsis, Marc Schoenauer, et al. "The Grand Challenge of Computer Go: Monte Carlo Tree Search and Extensions." *Communications of the ACM* 55, no. 3 (2012): 106–13.

31. Maddison, Chris J., Aja Huang, Ilya Sutskever, and David Silver. "Move Evaluation in Go Using Deep Convolutional Neural Networks." *arXiv*, April 10, 2015. http://arxiv.org/abs/1412.6564.

32. Silver, David, Aja Huang, Chris J. Maddison, et al. "Mastering the Game of Go with Deep Neural Networks and Tree Search." *Nature* 529, no. 7587 (January 2016): 484–89. https://doi.org/10.1038/nature16961.

33. Kuhn, H. W. "9. A Simplified Two-Person Poker." In *Contributions to the Theory of Games (AM-24)*, vol. I, edited by Harold William Kuhn and Albert William Tucker, 97–104. Princeton University Press, 1951. https://doi.org/10.1515/9781400881727-010.

34. Koller would later discover that someone *had* thought about it earlier, though only in passing. In a short (two-page) paper in the journal *Soviet Mathematics*, Romanovskii had proposed a sequence "quasistrategy" for games. Romanovskii's paper had no proofs, and there was no follow-up until the 1990s. By the '90s, computers were able to solve poker-sized sequence form programs. Sometimes the best ideas need to happen at the right time to be remembered. Romanovskii, I. V. "Reduction of a Game with Complete Memory to a Matrix Game." *Soviet Mathematics* 3, no. 3 (1962): 678–81.

35. For the technically minded, the number of pure strategies is typically exponential in the size of the game tree. Roughly speaking, its size is equal to the number of moves at any turn raised to the power of the number of nodes in the tree. Even if there are only three moves at any turn and 100 nodes, the number of pure strategies ends up around 10^{48}.

36. Koller, Daphne, Nimrod Megiddo, and Bernhard von Stengel. "Efficient Computation of Equilibria for Extensive Two-Person Games." *Games and Economic Behavior* 14, no. 2 (June 1996): 247–59. https://doi.org/10.1006/game.1996.0051.

37. Koller, Daphne, and Avi Pfeffer. "Representations and Solutions for Game-Theoretic Problems." In "Economic Principles of Multi-Agent Systems," special issue, *Artificial Intelligence* 94, no. 1 (July 1997): 167–215. https://doi.org/10.1016/S0004-3702(97)00023-4.

38. Other abstractions that were used were removing betting rounds, and considering pre-flop and post-flop betting as independent from each other.

39. Billings, D., N. Burch, A. Davidson, et al. "Approximating Game-Theoretic Optimal Strategies for Full-Scale Poker." IJCAI'03: *Proceedings of the 18th International Joint Conference on Artificial Intelligence*. August 2003. 661–68.

40. Zinkevich, Martin, Michael Johanson, Michael Bowling, and Carmelo Piccione. "Regret Minimization in Games with Incomplete Information." In *Advances*

in *Neural Information Processing Systems (NeurIPS)*, vol. 20. 2007. https://proceedings.neurips.cc/paper/2007/hash/08d98638c6fcd194a4b1e6992063e944-Abstract.html.

41. Bowling, Michael, Neil Burch, Michael Johanson, and Oskari Tammelin. "Heads-up Limit Hold'em Poker Is Solved." *Science* 347, no. 6218 (2015):145–49.

42. Romer, Keith. "How A.I. Conquered Poker." *The New York Times Magazine*, January 18, 2022. https://www.nytimes.com/2022/01/18/magazine/ai-technology-poker.html.

43. I could write a whole book about behavioral economics. But one of the most influential of such books, *Thinking Fast and Slow*, cites almost no reproducible studies. A detailed look at this issue can be found at https://replicationindex.com/2020/12/30/a-meta-scientific-perspective-on-thinking-fast-and-slow/

44. Williamson, Edwin, Caroline Soane, and J. Bryan Carmody. "The US Residency Match at 70: What Was, What Is, and What Could Be." *Journal of Graduate Medical Education* 14, no. 5: 519–21. https://doi.org/10.4300/JGME-D-22-00248.1.

45. Gale, D., and L. S. Shapley. "College Admissions and the Stability of Marriage." *American Mathematical Monthly* 69, no. 1 (March 1962): 9–15. https://doi.org/10.2307/2312726.

Chapter 4

1. Abraham, E. P., E. Chain, C. M. Fletcher, et al. "Further Observations on Penicillin." *The Lancet* 238, no. 6155 (August 16, 1941): 177–89. https://doi.org/10.1016/S0140-6736(00)72122-2.

2. Auyang, Sunny Y. *Engineering—An Endless Frontier*. Harvard University Press, 2006.

3. Vecchio, Ignazio, Cristina Tornali, Nicola Luigi Bragazzi, and Mariano Martini. "The Discovery of Insulin: An Important Milestone in the History of Medicine." *Frontiers in Endocrinology* 9 (October 2018): 613.

4. López-Muñoz, Francisco, Ronaldo Ucha-Udabe, and Cecilio Álamo. "The History of Barbiturates a Century after Their Clinical Introduction." *Neuropsychiatric Disease and Treatment* 1, no. 4 (December 2005): 329–43.

5. Emanuel, M. B. "Histamine and the Antiallergic Antihistamines: A History of Their Discoveries." *Clinical & Experimental Allergy* 29, no. S3 (July 1999): 1–11. https://doi.org/10.1046/j.1365-2222.1999.00004.x-i1.

6. Benedek, T. G. "History of the Development of Corticosteroid Therapy." *Clinical and Experimental Rheumatology* 29, suppl. 68 (2011): S5–S12.

7. Farber, Sidney, Louis K. Diamond, Robert D. Mercer, Robert F. Sylvester Jr., and James A. Wolff. "Temporary Remissions in Acute Leukemia in Children Produced by Folic Acid Antagonist . . . Aminopterin." *CA: A Cancer Journal for Clinicians* 24, no. 5 (1974): 297–305. https://doi.org/10.3322/canjclin.24.5.297.

8. Ballentine, Carol. "Sulfanilamide Disaster." *FDA Consumer*, June 1981. https://www.fda.gov/media/110479/download?attachment.

9. Sackett, David L., William M. C. Rosenberg, J. A. Muir Gray, R. Brian Haynes, and W. Scott Richardson. "Evidence Based Medicine: What It Is and What It Isn't." *BMJ* 312, no. 7023 (January 1996): 71–72. https://doi.org/10.1136/bmj.312.7023.71.

10. Crofton, J. "The MRC Randomized Trial of Streptomycin and Its Legacy: A View from the Clinical Front Line." *Journal of the Royal Society of Medicine* 99, no. 10 (October 2006): 531–34.

11. In this section, I'm calculating these odds using properties of the *binomial distribution*. The binomial distribution tells us the probability of seeing at most k heads in n coin flips. The formula for those who want to try this at home is

$$I_{\frac{1}{2}}(n-k, k+1)$$

where I is the *incomplete beta function*.

12. Every randomized experiment has some measurement error because we can only see what happens to a person if they get the treatment or not. We can't observe both outcomes where the same person is in treatment and control.

This argument is based on what is called a "*p*-value" in statistics, but the exact details of what a *p*-value means don't actually matter much for interpreting trial results. The calculation just gives us a way to interpret the results of the trial and to assess how much variation could come from our random assignment. *P*-values are only telling us about the errors from random assignment, not about anything else in an experiment. The least we can hope for in an experiment is that the gap between the outcomes in treatment and control should be large in your hypothetical replays of the experiment.

13. Bartlett, M. S. "Some Remarks on the Theory of Statistics." *Transactions of the Manchester Statistical Society* 18 (1951).

14. Hill, A. Bradford. "The Clinical Trial." *New England Journal of Medicine* 247, no. 4 (July 24, 1952): 113–19. https://doi.org/10.1056/NEJM195207242470401.

15. Hill, A. Bradford. "The Harben Lectures, 1957: The Experimental Approach in Preventive Medicine." *The Journal of the Royal Institute of Public Health and Hygiene* 21, no. 7 (1958): 185–96.

16. Medical Research Council. "The Prevention of Whooping-Cough by Vaccination." *British Medical Journal* 1, no. 4721 (June 30, 1951): 1463–71.

17. Meldrum, Marcia. "'A Calculated Risk': The Salk Polio Vaccine Field Trials of 1954." *British Medical Journal* 317, no. 7167 (October 31, 1998): 1233–36.

18. Paul, John Rodman. "Thomas Francis, Jr: July 15, 1900–October 1, 1969." In *National Academy of Sciences Biographical Memoirs*, vol. 44. United States, National Academies Press, 1974.

19. Francis, Thomas Jr. "Evaluation of the 1954 Poliomyelitis Vaccine Field Trial: Further Studies of Results Determining the Effectiveness of Poliomyelitis Vaccine

(Salk) in Preventing Paralytic Poliomyelitis." *Journal of the American Medical Association* 158, no. 14 (August 6, 1955): 1266–70. https://doi.org/10.1001/jama.1955.02960140028004.

20. In contemporary epidemiological parlance, we might say that the Salk vaccine was 71% *effective* against paralytic polio.

21. Marks, Harry M. "The 1954 Salk Poliomyelitis Vaccine Field Trial." *Clinical Trials* 8, no. 2 (April 8, 2011): 224–34. https://doi.org/10.1177/1740774511399110.

22. Meier, Paul. "Safety Testing of Poliomyelitis Vaccine." *Science* 125, no. 3257 (May 31, 1957): 1067–71. https://doi.org/10.1126/science.125.3257.1067.

23. Egan, Robert L. "Experience with Mammography in a Tumor Institution: Evaluation of 1,000 Studies." *Radiology* 75, no. 6 (December 1960): 894–900. https://doi.org/10.1148/75.6.894.

24. Shimkin, Michael B. "Cancer of the Breast: Some Old Facts and New Prospectives." *JAMA* 183, no. 5 (February 2, 1963): 358–61.

25. Shimkin, Michael B. "The Numerical Method in Therapeutic Medicine." *American Journal of Hospital Pharmacy* 21, no. 6 (June 1964): 277–85. https://doi.org/10.1093/ajhp/21.6.277.

26. Friedman, Adele K., Samuel I. Askovitz, Simon M. Berger, et al. "A Co-Operative Evaluation of Mammography in Seven Teaching Hospitals." *Radiology* 86, no. 5 (May 1966): 886–91. https://doi.org/10.1148/86.5.886; Stevens, G. Melvin, and John F. Weigen. "Mammography Survey for Breast Cancer Detection. A 2-Year Study of 1,223 Clinically Negative Asymptomatic Women over 40." *Cancer* 19, no. 1 (January 1966): 51–59. https://doi.org/10.1002/1097-0142(196601)19:1<51::AID-CNCR2820190105>3.0.CO;2-R.

27. Shimkin, M. B. "In the Middle: 1954—63—Historical Note." *Journal of the National Cancer Institute* 62, no. 5 (May 1979): 1295–317.

28. Nagourney, Eric. "Dr. Philip Strax, 90, Is Dead; An Advocate of Mammograms." *The New York Times*, March 11, 1999, 29. https://www.nytimes.com/1999/03/11/us/dr-philip-strax-90-is-dead-an-advocate-of-mammograms.html.

29. Technically speaking, you can do more sophisticated statistics to try to get at the direct effect of the mammogram, but these techniques are complex and difficult even for statisticians to understand. In this chapter, I'm focusing on what can be directly measured without additional statistical storytelling, though many econometricians will object to my anti-statistical position.

30. Shapiro, Sam. "Evidence on Screening for Breast Cancer from a Randomized Trial." *Cancer* 39, no. 6 (1977): 2772–82.

31. The number of people who hypothetically have to receive a treatment in order for one person to benefit is called the number needed to treat. This number is the inverse of the absolute risk reduction. For the HIP trial, 1/0.0007 is approximately 1400.

32. Strax, Philip. *Early Detection: Breast Cancer is Curable.* Harper & Row, 1974.

33. Bailar, John C. III. "Mammography: A Contrary View." *Annals of Internal Medicine* 84, no. 1 (January 1976): 77–84. https://doi.org/10.7326/0003-4819-84-1-77.

34. Shen, Yu, and Marvin Zelen. "Screening Sensitivity and Sojourn Time from Breast Cancer Early Detection Clinical Trials: Mammograms and Physical Examinations." *Journal of Clinical Oncology* 19, no. 15 (August 2001): 3490–99. https://doi.org/10.1200/JCO.2001.19.15.3490.

Current techniques have improved mammography by quite a large factor. Recent estimates suggest that mammograms can detect 85 percent of breast cancers. Still, even with these improvements, no randomized trial has found unequivocal benefit of screening programs. Pisano, Etta D., R. Edward Hendrick, Martin J. Yaffe, et al. "Diagnostic Accuracy of Digital versus Film Mammography: Exploratory Analysis of Selected Population Subgroups in DMIST. *Radiology* 246, no. 2 (February 2008): 376–83. https://doi.org/10.1148/radiol.2461070200.

35. Shimkin, "In the Middle."

36. Gøtzsche, Peter C., and Ole Olsen. "Is Screening for Breast Cancer with Mammography Justifiable?" *The Lancet* 355, no. 9198 (January 8, 2000): 129–34.

37. Shapiro, Sam, Wanda Venet, Philip Strax, and Louis Venet. *Periodic Screening for Breast Cancer: The Health Insurance Plan Project and Its Sequelae, 1963–1986*. Johns Hopkins University Press, 1988. 17.

38. Olsen, Ole, and Peter C. Gøtzsche. "Screening for Breast Cancer with Mammography. *Cochrane Database of Systematic Reviews* 2001, Issue 4: CD001877. https://doi.org/10.1002/14651858.CD001877.

39. Some of these plausibility arguments are presented in a 2004 review: Freedman, David A., Diana B. Petitti, and James M. Robins. "On the Efficacy of Screening for Breast Cancer." *International Journal of Epidemiology* 33, no. 1 (February 2004): 43–55. https://doi.org/10.1093/ije/dyg275.

40. Death classification is hard. Is assisted suicide a cancer death? Is pneumonia from a weakened immune system a cancer death? Trying to answer questions like this still troubles trialists today.

41. For example, the sampled accounting of how many women had prior breast lumps would not be allowed in a modern clinical trial.

42. Biller-Andorno, Nikola, and Peter Jüni. "Abolishing Mammography Screening Programs? A View from the Swiss Medical Board." *New England Journal of Medicine* 370, no. 21 (May 22, 2014): 1965–67. https://doi.org/10.1056/NEJMp1401875.

43. https://www.uspreventiveservicestaskforce.org/uspstf/draft-recommendation/breast-cancer-screening-adults.

44. Domenighetti, Gianfranco, Barbara D'Avanzo, Matthias Egger, et al. "Women's Perception of the Benefits of Mammography Screening: Population-Based Survey in Four Countries." *International Journal of Epidemiology* 32 (2003): 816–21. https://core.ac.uk/reader/212372912.

45. Sandhu, Gurdarshan S., and Gerald L. Andriole. "Overdiagnosis of Prostate Cancer." *Journal of the National Cancer Institute. Monographs* 2012, no. 45 (September 2012): 146–51. https://doi.org/10.1093/jncimonographs/lgs031.

46. Centers for Disease Control and Prevention. "An Update on Cancer Deaths in the United States." Atlanta, GA: US Department of Health and Human Services, Centers for Disease Control and Prevention, Division of Cancer Prevention and Control. 2022. https://stacks.cdc.gov/view/cdc/119728.

47. https://www.statnews.com/2020/10/27/colorectal-cancer-screening-should-start-five-years-earlier-at-45-expert-panel-says/.

48. Bretthauer, Michael, Magnus Løberg, Paulina Wieszczy, et al. "Effect of Colonoscopy Screening on Risks of Colorectal Cancer and Related Death." *New England Journal of Medicine* 387, no. 17 (October 9, 2022): 1547–56. https://doi.org/10.1056/NEJMoa2208375.

49. This dataset of study data in the Cochrane Database indexes over 30,000 clinical trials. Simon Schwab and Erik van Zwet. "Re-estimating 400,000 treatment effects from intervention studies in the Cochrane Database of Systematic Reviews." Open Science Framework, 2020. https://doi.org/10.17605/OSF.IO/XJV9G.

50. Echt, Debra. "CAST: A Study That Rocked the Cardiology World and Became the Poster Child for Evidence-Based Medicine." *Heart Rhythm* 21, no. 2 (February 2024): 131–32. https://doi.org/10.1016/j.hrthm.2023.09.030.

51. Manson, JoAnn E., Rowan T. Chlebowski, Marcia L. Stefanick, et al. "Menopausal Hormone Therapy and Health Outcomes During the Intervention and Extended Poststopping Phases of The Women's Health Initiative Randomized Trials." *JAMA: The Journal of the American Medical Association* 310, no. 13 (October 2, 2013): 1353–68. https://doi.org/10.1001/jama.2013.278040.

52. Manson, JoAnn E., Carolyn J. Crandall, Jacques E. Rossouw, et al. "The Women's Health Initiative Randomized Trials and Clinical Practice: A Review." *JAMA* 331, no. 20 (May 28, 2024): 1748–60. https://doi.org/10.1001/jama.2024.6542.

53. Angrist, Joshua D., and Jörn-Steffen Pischke. "The Credibility Revolution in Empirical Economics: How Better Research Design Is Taking the Con out of Econometrics." *Journal of Economic Perspectives* 24, no. 2 (Spring 2010): 3–30. https://doi.org/10.1257/jep.24.2.3.

54. Pryce, Joseph, Marty Richardson, and Lengeler C. "Insecticide-Treated Nets for Preventing Malaria." *Cochrane Database of Systematic Reviews* 2018, Issue 11 (2018): CD000363. http://doi.org/10.1002/14651858.CD000363.pub3.

55. Miguel, Edward, and Michael Kremer. "Worms: Identifying Impacts on Education and Health in the Presence of Treatment Externalities." *Econometrica* 72: 159–217. https://doi.org/10.1111/j.1468-0262.2004.00481.x.

56. Glewwe, Paul, Albert Park, and Meng Zhao. "A Better Vision for Development: Eyeglasses and Academic Performance in Rural Primary Schools in China."

Journal of Development Economics 122 (2016): 170–82. https://doi.org/10.1016/j.jdeveco.2016.05.007.

57. Miguel and Kremer, "Worms."

58. Aiken, Alexander M., Calum Davey, James R. Hargreaves, and Richard J. Hayes. "Re-analysis of Health and Educational Impacts of a School-Based Deworming Programme in Western Kenya: A Pure Replication." *International Journal of Epidemiology* 44, no. 5 (October 2015): 1572–80. https://doi.org/10.1093/ije/dyv127; Davey, Calum, Alexander M. Aiken, Richard J. Hayes, and James R. Hargreaves. "Re-analysis of Health and Educational Impacts of a School-Based Deworming Programme in Western Kenya: A Statistical Replication of a Cluster Quasi-Randomized Stepped-Wedge Trial." *International Journal of Epidemiology* 44, no. 5 (October 2015): 1581–92. https://doi.org/10.1093/ije/dyv128.

59. https://www.nytimes.com/2009/03/01/business/01marissa.html.

60. Muralidharan, Omkar, Niall Cardin, Todd Phillips, and Amir Najmi. "Causality in Machine Learning." The Unofficial Google Data Science Blog, January 31, 2017. https://www.unofficialgoogledatascience.com/2017/01/causality-in-machine-learning.html.

61. Lewis, Randall A., and Justin M. Rao. "The Unfavorable Economics of Measuring the Returns to Advertising." *The Quarterly Journal of Economics* 130, no. 4 (November 2015): 1941–73. https://doi.org/10.1093/qje/qjv023.

62. Vivalt, Eva. "How Much Can We Generalize from Impact Evaluations?" *Journal of the European Economic Association* 18, no. 6 (December 2020): 3045–89. https://doi.org/10.1093/jeea/jvaa019.

Chapter 5

1. Shannon, C. E. "A Mathematical Theory of Communication." *The Bell System Technical Journal* 27, no. 3 (July 1948): 379–423. https://doi.org/10.1002/j.1538-7305.1948.tb01338.x.

2. In modern machine learning lingo, this is the cross-entropy loss of the language model evaluated on all text. This is the metric still used today to evaluate language models like those built at OpenAI.

3. Shannon, C. E. "Prediction and Entropy of Printed English." *The Bell System Technical Journal* 30, no. 1 (January 1951): 50–64. https://doi.org/10.1002/j.1538-7305.1951.tb01366.x.

4. Valmeekam, Chandra Shekhara Kaushik, Krishna Narayanan, Dileep Kalathil, Jean-Francois Chamberland, and Srinivas Shakkottai. "LLMZip: Lossless Text Compression Using Large Language Models." *arXiv*, June 26, 2023. https://doi.org/10.48550/arXiv.2306.04050.

5. Brown, Tom, Benjamin Mann, Nick Ryder, et al. "Language Models Are Few-Shot Learners." *Advances in Neural Information Processing Systems* 33 (2020). https://proceedings.neurips.cc/paper_files/paper/2020/hash/1457c0d6bfcb4967418bfb8a c142f64a-Abstract.html.

6. Robbins, Herbert, and Sutton Monro. "A Stochastic Approximation Method." *Annals of Mathematical Statistics* 22, no. 3 (1951): 400–407. https://doi.org/10.1214/aoms/1177729586.

7. McCulloch, Warren S., and Walter Pitts. "A Logical Calculus of the Ideas Immanent in Nervous Activity." *Bulletin of Mathematical Biophysics* 5 (December 1943): 115–33. https://doi.org/10.1007/BF02478259.

8. In one of my favorite parodies of academic papers, David Fouhey implemented a deep neural network in Excel. You don't need that many primitive operations to simulate arbitrary computer programs. https://www.deepexcel.net/.

9. All these examples are real, publicly available data. If you'd like to try out some machine learning for yourself, their links are:

https://archive.ics.uci.edu/dataset/53/iris

https://archive.ics.uci.edu/dataset/17/breast+cancer+wisconsin+diagnostic

https://www.kaggle.com/code/prasadperera/the-boston-housing-dataset

10. Shannon, "A Mathematical Theory of Communication."

11. Yeang, C.-P. "Tubes, Randomness, and Brownian Motions: Or, How Engineers Learned to Start Worrying about Electronic Noise." *Archive for History of Exact Sciences* 65 (April 24, 2011): 437–70. https://doi.org/10.1007/s00407-011-0082-5.

12. Nyquist, H. "Thermal Agitation of Electric Charge in Conductors." *Physical Review* 32, no. 1 (July 1928): 110–13. https://doi.org/10.1103/PhysRev.32.110.

13. Wiener, Norbert. *Extrapolation, Interpolation, and Smoothing of Stationary Time Series: With Engineering Applications.* MIT Press, 1949. (open access at https://direct.mit.edu/books/oa-monograph/4361/Extrapolation-Interpolation-and-Smoothing-of).

14. Kalman, Rudolf E. "Randomness Reexamined." *MIC Journal: Modeling, Identification and Control* 15, no. 3 (1994): 141–51. https://doi.org/10.4173/mic.1994.3.3.

15. Statisticians have proposed other models, though they end up being more or less the same. For instance, a popular alternative to the iid model is called "exchangeability."

We say the sequence of events is *exchangeable* if the probability of the entire sequence is the same even if we shuffle the order of the events.

"Exchangeable" is not only saying that you can shuffle the past, but that you can also shuffle *the future* with the past and not be surprised. If I have seven dice colored like the rainbow, the odds of 1,2,3,4,3,2,1 are the same if I roll in the order red, orange, yellow, green, blue, indigo, violet, or if I roll in the order indigo, orange, violet, red, yellow, green, blue. And if I told you I was going to roll four today and three tomorrow, the odds of the sequence would be the same as if I rolled all seven today.

While any iid sequence is exchangeable, it is not true that any exchangeable distribution is iid. That said, for the purpose of this text, these fine-grained distinctions won't be important.

16. The material here on Highleyman is expanded from the section is adapted and expanded from a blog post co-authored by Moritz Hardt at the blog *argmin.net*. The post has also been adapted to be the section 'Highleyman's data' in Chapter 8 of the textbook *Patterns, Predictions, and Actions: Foundations of Machine Learning* by Moritz Hardt and Benjamin Recht (Princeton University Press, 2022). With permission from the authors, we include much of the original text here with modifications made in tone and context.

17. Feigenbaum, James, and Daniel P. Gross. "Answering the Call of Automation: How the Labor Market Adjusted to Mechanizing Telephone Operation." *The Quarterly Journal of Economics* 139, no. 3 (August 2024): 1879–939. https://doi.org/10.1093/qje/qjae005.

18. Highleyman, Bill. "The Availability Digest: Dr. Bill's Doctoral Thesis." https://availabilitydigest.com/public_articles/1208/thesis.pdf.

19. Highleyman, W. H., and L. A. Kamentsky. "A Generalized Scanner for Pattern- and Character-Recognition Studies." In *Papers Presented at the March 3–5, 1959, Western Joint Computer Conference*, 291–94. IRE-AIEE-ACM '59 (Western). New York, NY, USA: Association for Computing Machinery, 1959. https://doi.org/10.1145/1457838.1457894.

20. Highleyman, Wilbur Hull. *Linear Decision Functions, with Application to Pattern Recognition*. PhD Dissertation, Polytechnic Institute of Brooklyn, 1961.

21. Here's the Excel formula where row 2 has my relevant data, and I've stashed the weights in row 1000:

=SIGN(SUMPRODUCT(B2:W2, B1000:W1000))

22. This is a slight simplification of the HEART Score for major cardiac events, a risk assessment tool widely used in emergency medicine. Six, A. J., B. E. Backus, and J. C. Kelder. "Chest Pain in the Emergency Room: Value of the HEART Score." *Netherlands Heart Journal* 16, no. 6 (June 2008): 191–96. https://doi.org/10.1007/BF03086144.

23. Highleyman, W. H. "Linear Decision Functions, with Application to Pattern Recognition." In *Proceedings of the IRE* 50, no. 6 (June 1962): 1501–14. https://doi.org/10.1109/JRPROC.1962.288194.

24. Rosenblatt, Frank. "The Perceptron: A Probabilistic Model for Information Storage and Organization in the Brain." *Psychological Review* 65, no. 6 (November 1958): 386–408.

25. Novikoff, Albert B. J. "On Convergence Proofs on Perceptrons." In *Symposium on the Mathematical Theory of Automata*, 615–22. Interscience, 1962.

26. Highleyman, W. H. "The Design and Analysis of Pattern Recognition Experiments." *Bell System Technical Journal* 41, no. 2 (March 1962): 723–44. https://doi.org/10.1002/j.1538-7305.1962.tb02426.x.

27. Chervonenkis, Alexey. "Chervonenkis's Recollections." In *Measures of Complexity: Festschrift for Alexey Chervonenkis*, edited by Vladimir Vovk, Harris Papadopoulos, and Alexander Gammerman, chapter 1. Springer International Publishing, 2015. https://doi.org/10.1007/978-3-319-21852-6.

28. Highleyman was arguing that the error on the training set should go to zero at a reasonable rate because the errors were binomially distributed. But he failed to account for the multiple testing phenomenon when fitting such linear predictors. The dimension had to matter in some regard. Chervonenkis realized this would require a notion of *uniform convergence* of the probability estimate over all linear prediction functions.

29. Highleyman, W. H., and L. A. Kamentsky. "Comments on a Character Recognition Method of Bledsoe and Browning." *IRE Transactions on Electronic Computers* EC-9, no. 2 (June 1960): 263–263. https://doi.org/10.1109/TEC.1960.5219829.

30. Bledsoe, W. W. "Further Results on the N-tuple Pattern Recognition Method." *IRE Transactions on Electronic Computers* EC-10, no. 1 (March 1961): 96–96. https://doi.org/10.1109/TEC.1961.5219162.

31. Chow, C. K. "An Optimum Character Recognition System Using Decision Functions." *IRE Transactions on Electronic Computers* EC-6, no. 4 (December 1957): 247–54. https://doi.org/10.1109/TEC.1957.5222035.

32. Chow, C. K. "A Recognition Method Using Neighbor Dependence." *IRE Transactions on Electronic Computers* EC-11, no. 5 (October 1962): 683–90. https://doi.org/10.1109/TEC.1962.5219431.

33. Highleyman, W. H. "Data for Character Recognition Studies." *IEEE Transactions on Electronic Computers* EC-12, no. 2 (April 1963): 135–36. https://doi.org/10.1109/PGEC.1963.263427.

34. Calvert, T. W. "Nonorthogonal Projections for Feature Extraction in Pattern Recognition." *IEEE Transactions on Computers* C-19, no. 5 (May 1970): 447–52. https://doi.org/10.1109/T-C.1970.222943.

35. Knoll, A. L. "Experiments with 'Characteristic Loci' for Recognition of Handprinted Characters." *IEEE Transactions on Computers* C-18, no. 4 (April 1969): 366–72. https://doi.org/10.1109/T-C.1969.222664.

36. Smith, D. R. "A Partitioning Method for Combinational Synthesis." *IEEE Transactions on Computers* C-17, no. 1 (January 1968): 72–75. https://doi.org/10.1109/TC.1968.5008873.

37. Ullmann, Julian Richard. *An Algebraic Technique for the Automatic Recognition of Visual Patterns*. PhD Dissertation, Imperial College, University of London, 1968.

38. Zobrist, Albert L. "The Organization of Extracted Features for Pattern Recognition." *Pattern Recognition* 3, no. 1 (April 1971): 23–30. https://doi.org/10.1016/0031-3203(71)90004-5.

39. Munson, J. H., R. O. Duda, and P. E. Hart, "Experiments with Highleyman's Data." *IEEE Transactions on Computers* C-17, no. 4 (April 1968): 399–401. https://doi.org/10.1109/TC.1968.229391.

40. This is what we'd now call "one-versus-all classification." See Nilsson, Nils. J., *Learning Machines: Foundations of Trainable Pattern-Classifying Systems*. McGraw-Hill, 1965.

41. Michalski, R. S., J. G. Carbonell, and T. M. Mitchell, eds. *Machine Learning: An Artificial Intelligence Approach*. Springer-Verlag, 1983. https://doi.org/10.1007/978-3-662-12405-5.

42. Anderson, D., ed. "Neural Information Processing Systems 0 (NIPS 1987)." NeurIPS Proceedings. https://proceedings.neurips.cc/paper_files/paper/1987.

43. John Platt, personal communication. Email correspondence between October 16 and 18, 2020.

44. Quinlan, J. R. "Induction of Decision Trees." *Machine Learning* 1 (March 1986): 81–106. https://doi.org/10.1007/BF00116251.

45. https://archive.ics.uci.edu/.

46. Bottou, L., C. Cortes, J. S. Denker, et al. "Comparison of Classifier Methods: A Case Study in Handwritten Digit Recognition." In *Proceedings of the 12th IAPR International Conference on Pattern Recognition* (Jerusalem, Israel, 1994), *vol. 3, conference C: Signal Processing* (cat. no. 94CH3440–5). IEEE Computer Society Press, 1994. 77–82, v. 2. https://doi.org/10.1109/ICPR.1994.576879.

47. In statistics, heuristic reasoning suggests that the test error should be inversely proportional to the square root of the number of training examples. With this heuristic, we would expect an 11× improvement over Munson's approaches. The actual recorded improvement from 12% to 0.7% was closer to 17×, not far from what the back-of-the-envelope calculation predicts.

48. Cohen, G., S. Afshar, J. Tapson, and A. van Schaik. "EMNIST: Extending MNIST to Handwritten Letters," *2017 International Joint Conference on Neural Networks (IJCNN)*, Anchorage, AK, 2017, 2921–26. https://doi.org/10.1109/IJCNN.2017.7966217.

49. Image Classification on EMNIST-Balanced. paperswithcode.com/sota/image-classification-on-emnist-balanced.

50. Applying the same naive scaling argument as above, the increase in dataset size would predict a 7× improvement if such an improvement was achievable.

51. Deng, Jia, Wei Dong, Richard Socher, Li-Jia Li, Kai Li, and Li Fei-Fei. "ImageNet: A Large-Scale Hierarchical Image Database." *2009 IEEE Conference on Computer Vision and Pattern Recognition*, Miami, FL, 2009, 248–55. https://doi.org/10.1109/CVPR.2009.5206848.

52. Princeton University. "About WordNet." WordNet: A Lexical Database for English. 2010. https://wordnet.princeton.edu/.

53. https://www.flickr.com/.

54. https://www.mturk.com/.

55. Russakovsky, Olga, Jia Deng, Hao Su, et al. "ImageNet Large Scale Visual Recognition Challenge." *International Journal of Computer Vision* 115 (2015): 211–52. https://doi.org/10.1007/s11263-015-0816-y.

56. Bennett, James, and Stan Lanning. "The Netflix Prize." *Proceedings of KDD Cup and Workshop*, 2007.

57. Krizhevsky, Alex, Ilya Sutskever, and Geoffrey E. Hinton. "ImageNet Classification with Deep Convolutional Neural Networks." *Advances in Neural Information Processing Systems* 25 (2012): 1087–105.

58. Simon, Herbert A. "Why Should Machines Learn?" In *Machine Learning: An Artificial Intelligence Approach*, edited by R. S. Michalski, J. G. Carbonell, and T. M. Mitchell. Springer-Verlag, 1983. https://doi.org/10.1007/978-3-662-12405-5.

59. "New Navy Device Learns by Doing." *The New York Times*, July 8, 1958, 25.

60. Minsky, Marvin, and Seymour A. Papert. *Perceptrons: An Introduction to Computational Geometry.* Reissue of the 1988 expanded edition with a new foreword by Léon Bottou. MIT Press, 2017.

61. Shankar, V., R. Roelofs, H. Mania, A. Fang, B. Recht, and L. Schmidt. "Evaluating Machine Accuracy on ImageNet." *Proceedings of the 37th International Conference on Machine Learning. PMLR* 119 (2020): 8634–44. https://proceedings.mlr.press/v119/shankar20c.html.

62. Jumper, J., R. Evans, A. Pritzel, et al. "Highly Accurate Protein Structure Prediction with AlphaFold." *Nature* 596 (2021): 583–89. https://doi.org/10.1038/s41586-021-03819-2.

63. https://www.nobelprize.org/prizes/chemistry/2024/press-release/.

64. Strickland, Eliza. "IBM Watson, Heal Thyself: How IBM Overpromised and Underdelivered on AI Health Care." *IEEE Spectrum* 56, no. 4 (April 2019): 24–31. https://doi.org/10.1109/MSPEC.2019.8678513.

65. Wong, Andrew, Erkin Otles, John P. Donnelly, et al. "External Validation of a Widely Implemented Proprietary Sepsis Prediction Model in Hospitalized Patients." *JAMA Internal Medicine* 181, no. 8 (2021): 1065–70. https://doi.org/10.1001/jamainternmed.2021.2626.

Chapter 6

1. Flood, Merrill M. "Some Experimental Games." RAND Corporation Memorandum, 1952. https://www.rand.org/content/dam/rand/pubs/research_memoranda/2008/RM789-1.pdf.
2. Meehl, Paul E. "Causes and Effects of My Disturbing Little Book." *Journal of Personality Assessment* 50, no. 3 (September 1986): 370–75. https://doi.org/10.1207/s15327752jpa5003_6.
3. Meehl, Paul E. *Clinical Versus Statistical Prediction: A Theoretical Analysis and a Review of the Evidence.* University of Minnesota Press, 1954.
4. Though this book appeared in 1954, Meehl had assembled the main ideas in a lecture series he began delivering in 1947. After several years facing rejections from publishers, Meehl eventually managed to convince his own university's press to publish his manuscript.
5. Algorithmic recidivism prediction remains a contentious topic. It is one of the most popular examples discussed by the machine learning fairness community. The common refrain is to argue that these risk assessments are opaque and potentially biased. But Burgess was attempting to make the case for a more liberal parole system. He thought his algorithm could be less political, more fair, and more accurate.
6. Meehl, "Causes and Effects of My Disturbing Little Book."
7. Grove, William M., David H. Zald, Boyd S. Lebow, Beth E. Snitz, and Chad Nelson. "Clinical Versus Mechanical Prediction: A Meta-Analysis." *Psychological Assessment* 12, no. 1 (2000): 19–30. https://doi.org/10.1037/1040-3590.12.1.19; Ægisdóttir, Stefanía, Michael J. White, Paul M. Spengler, et al. "The Meta-Analysis of Clinical Judgment Project: Fifty-Six Years of Accumulated Research on Clinical Versus Statistical Prediction." *The Counseling Psychologist* 34, no. 3 (May 1, 2006): 341–82. https://doi.org/10.1177/0011000005285875.
8. https://www.nobelprize.org/prizes/economic-sciences/2002/kahneman/biographical/.
9. Gawande, Atul. "The Checklist." *The New Yorker*, December 2, 2007, 86–95. https://www.newyorker.com/magazine/2007/12/10/the-checklist.
10. Klein, Gary. *Streetlights and Shadows: Searching for the Keys to Adaptive Decision Making.* MIT Press, 2009.
11. These examples are highlighted on the blog Paleofuture: https://paleofuture.com/blog/2010/12/9/driverless-car-of-the-future-1957.html; https://paleofuture.com/blog/2007/5/11/disneys-magic-highway-usa-1958.html.
12. Pomerleau, D. A. "Alvinn: An Autonomous Land Vehicle in a Neural Network. In *Advances in Neural Information Processing Systems* 1 (NeurIPS 1988). A demo of the early prototype is on YouTube: https://www.youtube.com/watch?v=ntIczNQKfjQ.

13. https://www.darpa.mil/about-us/timeline/-grand-challenge-for-autonomous-vehicles.

14. https://www.youtube.com/watch?v=cdgQpa1pUUE.

15. Google Self-Driving Car Project Monthly Report, August 2015.

16. https://jalopnik.com/let-s-all-laugh-at-these-bad-autonomous-car-predictions-1846690460.

17. https://www.theguardian.com/technology/2016/jun/02/self-driving-car-elon-musk-tech-predictions-tesla-google.

18. https://www.cnbc.com/2019/04/22/elon-musk-says-tesla-robotaxis-will-hit-the-market-next-year.html.

19. https://www.mckinsey.com/industries/automotive-and-assembly/our-insights/mobilitys-future-an-investment-reality-check.

20. In 2024, Alphabet claimed that Waymo lost between 0.8 and 1.1 billion dollars per quarter. Dividing this number by the number of reported rides estimates a loss of $1,200–$1,600 per ride. https://www.cnbc.com/2024/07/23/alphabet-to-invest-5-billion-in-self-driving-car-unit-waymo.html.

21. It will be fun to see how long after this book's publication they become truly ubiquitous. I don't think it's impossible they'll figure out a way to scale and profit from self-driving cars, but it seems likely that unless they are afforded private roads devoid of other drivers, obstacles, people, or animals, that it's going to remain a costly, long slog to bring their autonomy up to the level promised well over a decade ago. And if you gave me 100 billion dollars, I could likely have built a few thousand miles of light rail. For perspective, San Francisco's existing light rail system, the MUNI only has 71 miles of track (see https://www.sfmta.com/getting-around/muni/muni-metro-light-rail). The entire New York subway system is less than 700 miles of track (https://new.mta.info/guides/riding-the-subway).

22. De Groot, A. D. *Thought and Choice in Chess.* The Hague: Mouton, 1978. (Original work published 1946).

23. Newell, Allen, J. C. Shaw, and H. A. Simon. "Chess-Playing Programs and the Problem of Complexity." In *Computer Games* I, edited by David N. L. Levy, 89–115. New York: Springer, 1988. https://doi.org/10.1007/978-1-4613-8716-9_7.

24. Chase, William G., and Herbert A. Simon. "The Mind's Eye in Chess." In *Visual Information Processing*, edited by William G. Chase, 215–81. Academic Press, 1973. https://doi.org/10.1016/B978-0-12-170150-5.50011-1. Newell, A., and H. A. Simon. *Human Problem Solving.* Prentice-Hall, 1972.

25. Klein, Gary. *Sources of Power: How People Make Decisions.* 20th Anniversary Edition. MIT Press, 2017.

26. Lopes, Lola L. "The Rhetoric of Irrationality." *Theory and Psychology* 1, no. 1 (1991): 65–82. https://doi.org/10.1177/0959354391011005.

27. Kahneman, Daniel, and Gary Klein. "Conditions for Intuitive Expertise: A Failure to Disagree." *American Psychologist* 64, no. 6 (2009): 515–26. https://doi.org/10.1037/a0016755.

28. I'm not using the terms here, but this distinction is inspired by the distinction between *techne* and *metis* made by James C. Scott in his 1998 book *Seeing Like a State* (Yale University Press). Scott's definitions are a bit more subtle and based in distinguishing between indigenous and state knowledge. His view of the state as a statistical computation engine closely aligns with the main themes in this book.

Chapter 7

1. Richard H. Thaler and Cass R. Sunstein. *Nudge: Improving Decisions about Health, Wealth, and Happiness.* Yale University Press, 2008.

2. Graeber, David. *The Utopia of Rules: On Technology, Stupidity, and the Secret Joys of Bureaucracy.* Penguin Random House, 2015.

3. Sarewitz, Daniel. "How Science Makes Environmental Controversies Worse." *Environmental Science & Policy* 7, no. 5 (October 2004): 385–403. https://doi.org/10.1016/j.envsci.2004.06.001.

4. Postman, Neil. *Technopoly: The Surrender of Culture to Technology.* Penguin Random House, 1992. 88.

5. Weizenbaum, Joseph. *Computer Power and Human Reason: From Judgment to Calculation.* W. H. Freeman and Company, 1976.

6. Weizenbaum, J. "How to Make a Computer Appear Intelligent." *Datamation* 8, no. 2 (February 1962): 24–26.

7. Weizenbaum, J. ELIZA—A Computer Program for the Study of Natural Language Communication Between Man and Machine. *Communications of the ACM* 9, no. 1 (1966): 36–45.

8. Polanyi, Michael. *The Tacit Dimension.* Anchor Books / Doubleday & Company, 1967. 3–4.

INDEX

Page numbers in *italics* refer to figures and tables.

abstractions, poker, 96
A/B testing: Google, 135, 136; randomized trials, 133–38
action bias, policymaking, 209
actuarial method, statistical prediction, 185–86
Advances in Neural Information Processing Systems (NeurIPS), 168, 169
Agricultural Research Service, USDA, 32
Aha, David, File Transfer Protocol (FTP), 168
AI. *See* Artificial Intelligence (AI)
Air Force, 79
airplanes, autonomous, 198
air traffic control systems, 198
algorithm(s): backpropagation, 50; decision systems, 211–12; gradient, 49; logistics, 46–47; local search, 49; prototype for checkers, 75–76; simulated annealing, 54–55
Alphabet, 196; on Waymo, 250n20
AlphaFold, 180
AlphaGo, 98; computation power assisting, 90
"Alpha-Zero" games, 98
Amazon, Mechanical Turk service, 171–72

American Cancer Society, on mammograms, 125
American Psychologist (journal), 204
aminopterin, discovery by Farber, 105
analytics in sports, phenomenon of, 4–5
anchoring heuristic, Kahneman and Tversky, 189
antibiotics, infection, 103–4
antihistamines, Benadryl, 104
anti-lock brake control system, 40, 41
Apollo Program, System 360–91, 50
app companies, software, 134–35
approximate value functions: scores, 73; Shannon's, 73–74
Army Air Forces, 39; Statistical Control Division, 20
Army Math Research Center, Madison, 236n14
Arrow, Kenneth, 79; game theory group, 79
Artificial Intelligence (AI): chat systems, 144; conception of, 99; Koller on game theory for, 90–91; machine learning as optimization research, 234n36; Meehl and AI systems, 201; nascent field of, 70; program Perceptron, 38
Artificially Intelligent chatbots, 9

artificial network models, 173–74
artificial neural net models, 176
artificial neural networks, McCulloch and Pitts proposing, 144
AT&T, 61–62
Aumann, Robert, game theory group, 79
Aurora, self-driving technology company, 195
automated decision rules, 192
automatic control, optimal control theory, 39–40
automation: decision making, 206; linking calculation and, 16
availability heuristic, Kahneman and Tversky, 188
aviation: control system, 60–61; safety, 198–99
axiomatization, Kolmogorov's, 12
axioms, program application, 34

backgammon, Berliner devising program for, 87–88
backpropagation, algorithm, 50
Bailar, John, on breast cancer detection, 125
Bankman-Fried, Sam, rationalizing fraud, 209
Bartlett, M. S., addressing Manchester Statistical Society, 110–11
basis method, Dantzig on, 25
behavioral economics, field, 183
Bell Labs: dataset of handwritten characters, 169–70; Highleyman collecting alphabets from, 162; optical character recognition experiments, 38; Shannon's language modeling at, 144
Bellman, Richard: games of timing, 80; game theory research, 79; invention of dynamic programming, 41–42; on optimal strategies of games, 70
Bellman's optimality principle, 41
Benadryl, antihistamines, 104
Berliner, Hans, program playing backgammon, 87–88
Bernstein, Alex: chess-playing machine, 74, 76; prototypes, 85–86
BFGS (Broyden, Fletcher, Goldfarb, and Shanno), 49
Big Data, "large scale" computing, 35
Biller-Andorno, Nikola, controversial mammography report, 129
BINAC, Northrop Aircraft Corporation, 232n15
binomial distribution, 239n11
biometry, statistical methods of, 105
Bitcoin, 57
blackjack, optimal strategy for winning, 6
Blackwell, David: games of timing, 80; game theory research, 79; on optimal strategies of games, 70; optimal strategy of hypothetical duel, 80–81
Bledsoe, Woody: intuition of, 170; method from Sandia Labs, 163–64
bounded-depth tree search, Shannon's, 73–74
Bowling, Mike, 96; leading poker efforts, 96–97; solving first "real" version of poker, 97
branches, game trees, 72
breast cancer: describing cause of death, 128; HIP Study, 122–24; mammographic screening for, 119–21; treatability of *symptomatic*, 130
Bryson, Arthur: algorithm, 63; analyzing motions of objects, 49–50
Bureau of Labor Statistics, 27, 29

Burgess, Ernest, Illinois Parole Board on recidivism, 186, 249n5
Burroughs Corporation, 164
Bush, George W., election between Gore and, 219
Bush, Vannevar: advancing defense logistics, 14; digital device development, 14–15; National Defense Research Committee (NDRC), 13

Cadence, 56
calculation, linking automation and, 16
calculus, 49
cancer: appearance on X-ray, 130; benefits of screening, 135–36; colonoscopy for colon screening, 131–32; death classification, 241n40; early detection, 120; HIP Study for breast screening, 122–24; popularity of early detection, 131; screening for breast cancer, 119–20. *See also* chemotherapy
Carnegie Mellon University (CMU), 165; automatic self-driving project vehicle, 194, 196
Carpenter, Rowena, on minimum cost recommendation, 28
CAST, randomized trial, 132
chance, future probability, 153–54
character recognition, Highleyman and Kamensky on project, 155–56
ChatGPT, 180; ELIZA and, 222; large language model, 178–79
checkers: computing optimal strategy, 87; Samuel programming computer to play, 75–78
chemical reactors, computers optimizing, 44

chemotherapy: breakthroughs in, 118–21; childhood leukemia, 112; innovations, 104–5
Chernobyl disaster of 1986, 59
Chervonenkis, Alexey, on Highleyman's empirical risk minimization method, 161–62
chess, 70, 90, 91; comparing zero-sum games, 70; computers teaching about, 99; experiments by de Groot, 199; game of perfect information, 67; Shannon on optimal strategy for, 71
Chinook project, 235n13
chip design, non-smooth optimization, 53–54
Chow, Chao Kong "C. K.," decision theory for pattern recognition, 164, 165
climate science, Sarewitz's focus, 218
clinical trials, randomized clinical trials (RCTs)
Clinical versus Statistical Prediction (Meehl), 185
Cochrane Library, 132
code book, 99
coin flip, predicting, 11–12
Cold war, 39, 51, 60; control systems, 40; optimization of computer, 58; rise of administrative state, 17
colon cancer, colonoscopy for screening, 131–32
colonoscopy, cancer screening, 131–32
common task, pattern recognition, 166
communication: randomness in, 150–51; removing noise, 151
communications theory, natural language processing, 140–42
Compaq, 56
computational gameplay, 99

computations, Bush's differential analyzer for, 14
computer(s): character recognition using, 155–56; computer chips, 34, 58; creation and scaling of modern, 22; decision making of humans and, 9–10; decision systems of humans and, 210–11; growth in, intelligence, 18; as ideal rational agent, 210; Koller on, for making decisions, 91; "large scale" for, 35; optimization algorithms and, 57; power of systems, 18; rationality of, 7; rules and, 217; technocratic bureaucracies, 212–13
computer-aided design, impact on computing industry, 56–57
Computer Power and Human Reason (Weizenbaum), 221
computer programming, term, 47
computing speed, 50–51
conditional entropy, prediction errors, 143–44
confidence interval, statistics, 239n12
control systems: aviation, 60–61; feedback principle, 40; modern car, 40; nuclear power plant, 60, 61; policy optimization, 40–41; Stein on designing, 58–60
counterfactual regret minimization (CFR), Zinkevich inventing, 97
count-to-chance conversion, 153
COVID-19 pandemic, campaign to "follow the science," 209
creativity: play and, 217; system design, 35
Cruise, GM, 196
cruise control, modern car, 40

Curry, Steph, probability of free throw, 153
Cutter Laboratories, 118

Dantzig, George: algorithm as "the simplex method," 25, 29; analyst at the Statistical Control Division, 20–21; applying simplex method to Stigler's diet problem, 29–31; on the "basis method," 25, 25–26; computing to solve linear programs, 36–37; dynamics of simplex method, 26–27, 27; linear programming, 24; on mathematical optimization, 22–23; Montalbano and, 36; on programming, 21–22; RAND Corporation, 39; simplex method, 38, 43, 48
DARPA. *See* US Defense Advanced Research Projects Agency (DARPA)
DARPA (Defense Advanced Research Projects Agency) program, 168
decision making: computers and humans, 9–10, 210–11; Klein on general theory of, 200–201; mathematical rationality, 18–19; participatory, 224–25; randomized trial as "gold standard" of, 105–6; rational, 18; Sarewitz on, 219–20; Weizenbaum on decisions and choices, 221–24
Deep Blue (IBM): chess-playing computer defeating Kasparov, 87; optimal Go strategy and, 90
DeepMind, startup company, 90
DeepMind's "AlphaFold," Google, 178
deep neural networks, 173
defense logistics, Bush advancing, 14

de Groot, Adriaan, chess experiments, 199–200
Department of Agriculture, 27; nutrient content from, 29
deterministic, blue of tiny random particles as, 11
deworming pill, randomized study in Kenya, 133–34
diabetes, insulin, 104
diet: linear constraints of nutrients, 24–25; optimizing simple, 25
diet planning: dynamics of simplex method, 26–27, 27; oversimplified version of problem, 25–26
differential analyzer, Bush's nascent computer as, 14
digital computers, linear programming for, 36
digital devices, development for military-driven calculations, 14–15
digital logic gates, computers with, 14
Douglas Aircraft Company, 39, 79
Dresher, Melvin, experiments on Prisoner's Dilemma, 83, 183
Duda, Richard: intuition of, 170; method of "1-nearest neighbor," 165; readability of Highleyman's characters, 166
duel theory, theory of games of timing, 81
dynamic programming: complex nonlinear system, 44–46; invention by Bellman at RAND Corporation, 41–42; invention of, 41
dynamic world, rules in, 212

Early Detection (Strax), 124
ECAD, 56
Econometric Society, 33

edge cases, driving model, 197
Effective Altruism, 208–9
Egan, Robert, radiology for breast screening, 119–21
Electronics (magazine), 51, 53
ELIZA, chatbot, 221–22
empirical risk minimization, Highleyman's method, 162
English, approximations for, 141–43
ENIAC computer, 15, 232n15
Enlightenment science, 13
Epic Systems, electronic health records, 180
epidemiology, polio, 116
Equal Protection Clause, Fourteenth Amendment, 219
estrogen therapy, Women's Health Initiative Trial, 132
events, independent and identically distributed, 154
exchangeability, 154, 244n15
exchangeable, sequence of events, 154, 244–45n15
expertise, humans defining, 205
eyeglasses example, 137; outcome in, 134

Fairchild Semiconductor, 51
Farber, Sidney: chemotherapy for leukemia, 112; discovery of aminopterin, 105
FDA, randomized trials, 132, 137–38
Federal Food, Drug, and Cosmetic Act in 1938, 105
Fermat, Pierre de, on odds in games of chance, 11
File Transfer Protocol (FTP), 168
financial crisis, subprime mortgage, 208
Fincher, David, ads on amazing technologies, 61–62

firefighters, Klein's work on, 200
FiveThirtyEight website, Silver, 4
Fleming, Alexander, penicillin discovery, 102
Fletcher's variable metric method, 49
Flickr, search query, 171
floating point operations, 10 billion, per second, 87
floating point unit, IBM adding, 38
Flood, Merrill, experiments on Prisoner's Dilemma, 82, 83, 183
Florey, H. W., penicillin administration, 102–4
folk theorem, game theory, 83
football, 153
FORTRAN language, 233n20; IBM developing, 47
Fouhey, David, deep neural network in Excel, 244n8
Fourteenth Amendment, Equal Protection Clause, 219
Francis, Thomas, testing vaccine efficacy, 116–17
free throws, 153
Freudian psychoanalysis, 185
fuel injection system, 40, 41

Gale, David, National Resident Matching Program (NRMP), 101
games: computation, 69; human interaction through, 65–66; normal form of, 91–92; one-player, 66; rules and, 217–18; sequence "quasistrategy" for, 237n34; tension between rules and play, 216–18; two-player, 66
games of timing, Bellman and Blackwell, 80
game theory, 15; Koller, 94–95; Koller for artificial intelligence, 90–91; mathematical rationality, 8;

optimal decision making, 182; pillar of, 17; predicting human behavior, 98–99; prescriptive model of behavior, 100; Prisoner's Dilemma, 82–83; research at RAND Corporation, 79–80; Second World War, 64; von Neumann and Morgenstern, 64–72
game trees: branches, 72; Go's, 88–89; leaf nodes, 72, 73; Monte Carlo tree search (MCTS), 89; nodes, 72; von Neumann and Morgenstern, 72
Georgian bean pie, *lobiano*, 232n11
Go (game), 91, 99; baby version "Gomuku," 221; computers conquering, 88–90; MCTS mastering 9x9 game, 89
goals, rationality, 3–4
Google: A/B testing, 136; A/B tests, 135; DeepMind's "AlphaFold," 178; self-driving car division, 196; self-driving car project, 195
Gore, Al, election between Bush and, 219
Gøtzsche, Peter, systematic review on mammography, 126–29
gradient, perturbation, 49
gradient descent, algorithm, 159
Graeber, David: on game playing, 214–16; on play, 215–16; tension between rules and play, 216–18
Great Depression, 28, 30
grocery budgeting example: goals and constraints, 23–24; optimization problem, 23
grocery budgeting problem, as linear program, 24

handwritten characters: Bell Labs releasing dataset of, 169–70;

corresponding to letters, 177–78; Highleyman's, 163
Hart, Peter: intuition of, 170; method of "1-nearest neighbor," 165; readability of Highleyman's characters, 166
Harvard, 75
HB. *See* "Heuristics and Biases" school (HB)
healthcare, checklists, rules, and guidelines, 191–92
Health Insurance Plan of Greater New York, 122
Health Insurance Plan Study (HIP Study): number needed to treat, 240n31; perception of mammography's effectiveness, 130–31; skepticism of, 125; Strax leading, 122–24; systematic review, 126–29
Hebrew University of Jerusalem, 188
heuristics and biases, Kahneman and Tversky, 188
"Heuristics and Biases school (HB), 183–84, 187, 188, 189, 193; anchoring heuristic, 189; availability heuristic, 188; heuristics and biases, 188–89; naturalistic decision making (NDM) and, 216; representativeness heuristic, 189; research, 202–3
Highleyman, Bill: algorithm predicting dataset, 159; data communication and transmission, 167; digitized alphabets, *164*; on errors, 246n28; experiments gauging readability of characters, 166; handwriting examples, *163*; Kamentsky and, 156; linear predictors, 157–58; optical character recognition, 155–56; optimization methods improving linear predictor, 158–59; pattern recognition, 167; prediction function minimizing mistakes, 160–62; schemes for pattern recognition, 162–65
Hill, Bradford: on clinical trial, 110; lectures on clinical trials, 111–12; on outcomes and treatments, 112–13; on "overall" health of general population, 113; on potential biases, 117; on randomized clinical trial (RCTs), 132; streptomycin clinical trial, 106; on vaccine trials, 114–16
HIP Study. *See* Health Insurance Plan Study (HIP Study)
Honeywell, 58, 165
horseplay, 216
Hotelling, Harold: on nonlinear control, 33, 42; term nonlinear, 48
Houston Mission Control Center, 50
human choice, Weizenbaum on decisions and, 221–24, 225
humans, decision making of computers and, 9–10
hyperplane, name, 158

IBM: business calculators, 37–38; chess-playing machine, 74; computing speed, 50–51; Deep Blue defeating Kasparov, 87; floating point unit, 38; Kalman and Koepcke, 44–46; mechanical punch calculators, 36; 704, 38–39, 50, 74, 86; 701, 76; 7030, 50; 7094, 38–39; stored-memory computer, 38; transistor-based supercomputer, 47
IBM Watson, *Jeopardy* (game show), 180
ideal rational agent: computers as, 210; unknown and random, 16–17

IEEE Conference on Decision and
Control (1989), 58
ImageNet, Li's team, 171–73
imperfect information, poker, 67–68
Imperial College, 165
incomplete beta function, binomial
distribution, 239n11
independent and identically
distributed (iid model), 154,
244n15, 244–45n15
Independent Electric Light and Power
Companies, advertisement by,
193–94
infection control protocols, Pronovost
on, 191
infections, penicillin for, 102–4
information age, 19, 140; trend toward
quantification, 211
information theory, natural language
processing, 140–42
Institute for Advanced Study, 75
Institute of Home Economics, 32
insulin, discovery and production, 104
integer program, 232n10
integrated circuits: Moore's prediction
on, 51–53; planning out, 53;
standard cells for design, 54
Intel: adopting SDA's tools, 56; Moore
as co-founder, 18; processor model
number 80386, 56
"intelligence," growth in computer, 18
International Conference on
Machine Learning (ICML), 168,
169, 174, 175
iPhone 14, 176
irrational, 183
irrationality, 2
irrational person, 3
Israel Defense Forces (IDF), 187, 188
iterative linearization, practice of, 46

Jeopardy (game show), IBM Watson
on, 180
Jüni, Peter, controversial mammography report, 129

Kafka, Franz, on bureaucracies, 212
Kahneman, Daniel: on evaluating
expertise, 205; predicting
performance, 187–88; Tversky and,
188–90, 202
Kalman, Rudolf: growing power of
computers, 43–44; Koepcke and,
44–46; optimal policy for dynamic
programming, 42–43; pioneer of
automatic control, 152
Kalman filter, invention, 43
Kamentsky, Louis: character
recognition project, 156; Highleyman and, 162–63
Kasparov, Garry, Deep Blue
defeating, 87
Kesler, Carl, version of Rosenblatt's
perceptron, 165
Klein, Gary: on decision fatigue,
211–12; on decisions under extreme
pressure, 200; on evaluating
expertise, 205; firefighter study, 200;
general theory of decision making,
200–201; on human decision-
making, 192
knowledge: human and robotic,
206–7; as justified truth belief, 3
Kocsis, Levente, invention of
Monte Carlo tree search
(MCTS), 89
Koepcke, R. W., Kalman and,
44–46
Koller, Daphne: on game theory for
artificial intelligence, 90–91;
Megiddo and, on strategies from

INDEX 261

sequences, 93–94; work on sequence forming games, 94–95
Kolmogorov, Andrey, on rigorous theory, 12
Krizhevsky, Alex, code for networking graphics cards, 173
Kubrick, Stanley, *2001: A Space Odyssey*, 39
Kuhn, Harold: game theory group, 79; simplified version of poker, 92–93

Lancet, The (journal), 102, 126
language model(s): prediction software, 147–48; randomness in communication, 150–51
languages: FORTRAN, 47, 233n20; programming, 47
Large Language Models, 147; ChatGPT, 178–79
leaf nodes, game trees, 72
Leonard, Robert, game metaphor shaping culture at RAND, 85
leukemia: discovery of aminopterin, 105; Farber's chemotherapy for childhood, 112
Levy, David, chess bet with McCarthy, 86
Lewis, Michael, statistical management of Oakland Athletics, 5
Lewis, Randall, A/B test for advertising campaigns, 137
Li, Fei-Fei, database ImageNet, 171–73
life, risk management, 1–2
Limit Texas Hold'em: poker game, 95–96; solving, 97
linear, term, 48
linearization, optimal control problems, 46

linear predictors, Highleyman on, 157–58
linear programming: Dantzig on, 24; grocery budgeting problem, 24; nutrients of diet, 24–25, 25
linear programs, Dantzig on computing to solve, 36–37
lobiano, Georgian bean pie, 232n11
local optimization, nonlinear programming, 48
logical rationality, randomized trial, 133
Lopes, Lola, on heuristic and biases experiments, 202–3
Lyft ride-sharing company, 195

McCarthy, John: on learning machine, 167; term "artificial intelligence," 86
McCulloch, Warren, artificial neural networks, 144
machine, demand for quantitative metrics, 168
machine learning: art of, 147; current understanding of, 180, 181; empirical risk minimization method, 162; field, 168; framing prototypical problem, 145–48; language model, 243n2; McCarthy and Simon on, 167; Meehl's book arguing for, 185; pattern classification problem, 178; predicting protein structure, 177, 178; Samuel's program 78; Shannon's language prediction game, 147; Shannon's methodology for language models, 145; statistical, 182; statistical prediction, 8; University of Toronto, 173–74
Machine Learning for Healthcare, 180
machine-readable, meaning of, 190

"Magic Highways," Disneyland TV
    program, 194
malaria, bed nets for preventing, 137
mammography: advocacy for, 126;
    American Cancer Society on, 125;
    breast cancer screening, 119–21;
    HIP Study, 122–24, 125, 130–31;
    principle of early detection, 120;
    statistics in, 240n29; Swiss Medical
    Board recommendation, 129–30;
    systemic review of, 126–29
Manchester Statistical Society, 110–11
Markov processes, Shannon on, 141
matching pennies, 70; 2 x 2 tables of,
    73; comparing zero-sum games, 70;
    simple game of, 68, 68–69
mathematical formulae, optimal
    control, 41
mathematical irrationality, decision
    making, 205–6
mathematically rational agent,
    creation of, 11
mathematical optimization: Dantzig's,
    22–23; grocery budgeting problem,
    27; modeling and maximizing, 7–8;
    pillar of, 17
mathematical probability, invention
    of, 11
mathematical rationality: decision
    making and, 10, 224–25; definition,
    4, 6; game theory, 8; human
    decision making, 182; Kahneman
    and Tversky's work, 189–90;
    mathematical optimization, 7–8;
    pillars of, 7–9, 17, 210; Pinker's view
    of, 4; power of, 17–18; randomized
    experiments, 8; Silver as defender
    of, 4; statistical prediction, 8;
    Weizenbaum on harnessing, 223–24

Mathematical Tables Project, 30, 31,
    35; Dantzig, 30; solving diet
    problem, 30
mathematic equations, constraints
    describing system, 41
mathematicians: linking calculation
    and automation, 16; rebuilding
    efforts after the war, 15–16; war
    effort contributions, 15
mathematics: codified, of 1930s, 16;
    utility and strategy, 65
mechanical guidelines, medical
    decision making, 191
Mechanical Turk, Amazon's, 171–72
Meehl, Paul: actuarial method, 185–86;
    Clinical versus Statistical Prediction,
    185; on conflict for soul of
    psychology, 184; mechanical
    decision making, 190; on metric
    chasing, 205; on quantification of
    "better," 203–4
Megiddo, Nimrod, Koller and, on
    strategies from sequences, 93–94
Microsoft Excel, 146–47. *See also*
    spreadsheet metaphor
military-industrial complex, 50
"minimum cost diet problem," in
    optimization classes, 28
Minsky, Marvin: attack on perceptron,
    176; machine learning as optimiza-
    tion research, 234n36
MIT (Massachusetts Institute of
    Technology), 62, 74
MNIST dataset, 170
Moneyball (film), 5
Monro, Sutton, invention of stochastic
    gradient descent, 144
Montalbano, Michael, Dantzig
    and, 36

Monte Carlo tree search (MCTS), invention of, 89–90
Moon, putting men on the, 50–51, 86
Moore, Gordon: on computing industry, 18; on integrated circuits, 58; on integrated circuits in computers, 51–53; optimism of, 86; predictions, 57, 61
Moore's Law, 194
Morgenstern, Oskar: on chess, 90; formulation of games, 214–15; theory of games, 78; on theory of games and economics, 64–72
MOSAICO, 55
Munson, John: on heuristic reasoning, 247n47; intuition of, 170; method of "1-nearest neighbor," 165; readability of Highleyman's characters, 166
Musk, Elon, on autonomous driving, 195

Nash, John, game theory group, 79
National Bureau of Standards (NBS), 30, 36. *See also* NIST
National Cancer Institute (NCI), 119; Biometry and Epidemiology Branch, 119
National Defense Research Committee (NDRC), 13
National Institute of Standards and Technology (NIST), 169, 170
National Medal of Science, Blackwell, 79
National Research Council, 28; recommended daily allowances of nutrients, 28
National Resident Matching Program (NRMP), 100–101

"Naturalistic Decision Making" (NDM), 184, 192, 199, 205; heuristics and biases (HB) and, 216
natural language processing, Shannon, 140–42
NDM. *See* "Naturalistic Decision Making" (NDM)
Nealy, Robert, Samuel's checker program and, 77
Nelder-Mead method, 49
neoclassical economics, rationality, 208
Netflix dataset, 172
Newell, Allen: decision making in chess, 199; on human irrationality, 202
Newton's laws of motion, 42
New York Times, The (newspaper), 62, 63, 175
NIST, formerly National Bureau of Standards (NBS), 30
Nobel Prize in Chemistry, DeepMind's "AlphaFold," 178
Nobel Prize in economics, Stigler, 32
nodes, game trees, 72
noise: communication, 150–51; removing, 151
nonlinear programming, 48
nonlinear systems: approximating curve with line, 45, 45–46; dynamic programming for, 44–46
non-smooth optimization, chip design, 53–54
normal form, game, 91–92
normative decision theory, mathematical rationality as, 7
Northrop Aircraft Corporation, BINAC machine, 232n15
Novikoff, Al, on private language of perceptron workers, 158

nuclear power plants, control system, 60, 61
Nudge (Thaler and Sunstein), 213
nudge politics, 213–14
null hypothesis, thought experiment, 108–9
nutrients, linear constraints, 24–25
nutrition science, Mediterranean diet, 1
NVIDIA, 57

Oakland Athletics, Lewis on statistical management of, 5
Olsen, Greg: ideas about rationality, 6; on rationality, 208; on role of analytics in football, 5
Olsen, Ole, systematic review on mammography, 126–29
one-player games, 66
On the Edge (Silver), 4
OpenAI's ChatGPT, 144
optimal control theory, 39; automatic control, 39–40
optimal decisions, statistical analyses for, 6–7
optimal diet, cost of, 30
optimization: goal of, 23; modeling language of, 41
optimization algorithms, computers and, 57
optimization model, 33
outcomes, value of, 17
Ozempic, weight loss probability, 153

Papert, Seymour, attack on perceptron, 176
parity function, perceptron, 176
Pascal, Blaise, on odds in games of chance, 11

Pat McAfee Show (ESPN show), 5
pattern classification, solvability, 178
pattern recognition: common task, 166; current understanding of, *180*, 181; Highleyman's approach, 160, 167; human manipulating, 201–2; Shannon on possible, 148–52; spreadsheet metaphor, 154–55
PC clones, 56
penicillin: administration for bacterial infections, 102–4; eye infections, 103; manufacture of, 104; response to treatment, 112
penny. *See* matching pennies
Perceptron: first artificial intelligence program, 38; parity function, 176; Rosenblatt's classification scheme, 158; system, 174–76
Perceptrons (Minsky and Papert), 176, 177
perfect information: chess as game of, 67; Go as a game of, 88–89
Pfeiffer, Avi, code of Koller and, for poker, 94
physicists, modeling microscopic world, 11
Pinker, Steven: assumption on computers, 9; on rationality, 3, 6, 19, 208
Pitt, Brat, *Moneyball* (film), 5
Pitts, Walter, artificial neural networks, 144
placebo-controlled trial, eliminating bias, 115
play, Graeber on, 215–16
PlayStation 5, power of, 87
poker, 70; abstractions, 96; comparing zero-sum games, 70; game with imperfect information, 67–68; Koller and Pfeiffer, 94; Kuhn inventing

simplified version, 92–93; solving Limit Hold'em, 99–100
poker-playing robot, Schaeffer leading team, 95–96
Polanyi, Michael, Weizenbaum on, 223
policymaking, action bias, 209
policy problems, formulation versus solution, 30
polio: epidemiology, 116; paralytic and nonparalytic, 117; Salk's vaccine, 116–18; vaccine trials, 116–18
polling, 1
Polyak's Heavy Ball Method, 49
Pontryagin, Lev, advances in optimal control application, 43
Postman, Neil: on expert knowledge, 220–21; on technology changing the meaning of words, 10
prediction errors, Shannon on, 143–44
preventive medicine, 113–18; vaccine trials, 114–18
Prisoner's Dilemma, 92; Flood and Dresher's experiments, 183; Flood's experiments on, 82, 83; game theory of, 82–83; 2x2 tables, 92
probabilistic thinking, 12
probability, formalization of, 11
programming, Dantzig's, 21–22
programming languages, FORTRAN, 47, 233n20
Pronovost, Peter, on safety checklists, 191, 192
protein structure, machine learning to predict, 177, 178
punch cards, 169; character recognition, 155
"Pygmalion" (Shaw), 221
Python code, parity of bit string, 176

quantum mechanics, von Neumann of, 15
Quinlan, Ross, decision tree introduction algorithms, 168

radiology, screening for breast cancer, 119–20
rain, probability, 153
RAND Corporation, 39, 41, 101, 133, 187, 236n14; founding of, 79; game theory research, 79–80
randomized clinical trials (RCTs): A/B test, 133–38; CAST, 132; credibility, 133; FDA, 132, 137–38; as "gold standard" for causal decision-making, 105–6; Hill's lectures on, 111–12; null hypothesis, 108–9; Salk vaccine, 118; statistical significance, 109; streptomycin, 106–10
randomized experiments: mathematical rationality, 8; pillar of, 17
randomness, 69; communication, 150–51; use of, 13
random processes, 12
Rao, Justin, A/B test for advertising campaigns, 137
rational choice theory, economists, 7
rational decision making, computer design and, 16
rationality, 2; of computers, 7; conception, 2; definition, 231n2; faith in, 217; notion of, 2; Pinker defining, 3–4; rules and, 216; word, 2–3
Rationality (Pinker), 3
recidivism, Burgess on, 186, 249n5
recognition-primed decision (RPD) model, Klein's, 201

register transfer level (RTL), high-level description of circuits, 55
representativeness heuristic, Kahneman and Tversky, 189
rheumatic fever, outcomes and treatment, 112
rheumatoid arthritis, steroids, 104
rigorous theory, Kolmogorov on, 12
risk analysis, chances and costs, 1
risk management, guiding lives, 1–2
Robins, Herbert, invention of stochastic gradient descent, 144
"Robinson Crusoe" economics, 66
robotic knowledge, human and, 206–7
"Rock, Paper, Scissors," 70, 92; 3 x 3 tables of, 73; comparing zero-sum games, 70; outcome table, 67; simplified outcome table for, 92; tournaments, 82; two-person game, 66
Rosenblatt, Frank: algorithm predicting dataset, 159; IBM 704, 159; perceptron algorithm, 165; perceptron classification scheme, 158–59; on work of 1950s, 167
Rosenblatt's perceptron algorithm, 63
Royal Institute of Public Health and Hygiene, 114

safety, quantifying "good driver," 197
safety checklists, Pronovost, 191
Salk, Jonas, vaccine trial, 116–18
Samuel, Arthur: algorithm, 235n11; focus on machines, 91; machine learning scheme, 88; "players" Alpha and Beta, 76–77; on programming computer to play checkers, 75–78; prototypes, 85–86; research at IBM, 76, 77; scientific computing effort, 74–75
Samuelson, Paul, on random nature of matching pennies, 69
Sandia Labs, Bledsoe's method, 163
Sangiovanni-Vincentelli, Alberto: impact of computer-assisted design, 56–57; mathematical optimization and search techniques, 54, 55
Sarewitz, Daniel: on Bush-Gore election (2000), 219; on illusion of authority, 218–21; on limits of technocratic knowledge, 220–21
Schaeffer, Jonathan, team building poker-playing robot, 95–96
Schiffman, Max, optimal strategy of hypothetical duel, 80–81
Schatz, Albert, streptomycin discovery, 106
science, Sarewitz's focus, 218–19
Scientific American (magazine), 74
SDA, founding, 56
SEAC, 37
Second World War, game theory, 64
Sedol, Lee, Go master game, 90
selection bias, problem-solving, 34
self-driving cars, 198; advertisement, 193–94; Carnegie Mellon University (CMU) project, 194, 196; complexity, 207; Google funded, 195; Stanford project, 195
Selleck, Tom, narrating TV commercials, 61–62
sequences: Koller and Megiddo's strategies of, 93–94; Koller's work on, forming games, 94–95; "quasistrategy" for games, 237n34
Shannon, Claude: algorithm called stochastic gradient descent, 144;

approximate value functions, 73–74; bounded-depth tree search, 73–74; on building chess-playing robot, 99; on computer chess, 75, 87; on computer chip architectures, 14; focus on machines, 91; language modeling at Bell Labs, 144, 145; language prediction game, 147; machine learning, 145; Markov processes, 141; mathematical activity of, 15; modeling randomness, 152; monologue in documentary *Tomorrow*, 62–63; natural language processing, 140–42; on optimal strategy for chess, 71; possible pattern recognition, 148–52; predictable signals, 141; prediction errors, 143–44; simulation for approximations of English, 141–43; tree search, 199; on use of game trees, 72

Shapley, Lloyd: game theory research, 79; National Resident Matching Program (NRMP), 101

Shaw, Cliff: decision making in chess, 199; on human irrationality, 202

Shaw, George Bernard, "Pygmalion," 221

Shimkin, Michael: on colon cancer screening, 131–32; on mammographic screening for breast cancer, 119–21, 125

Silver, Nate: assumption on computers, 9; election analyst, 4; gaming analogy, 4–5; ideas about rationality, 6; on mathematical rationality, 19; on rationality, 208

Simon, Herbert: decision making in chess, 199; on human irrationality, 202; on learning machine, 167; on past, present, and future of machine learning, 174–76; on von Neumann and Morgenstern formulations, 78

"simplex method": Dantzig's algorithm as, 25, 29; dynamics of, in diet planning, 26–27, 27; invention of, 46

simplex technique, term, 232n7

simulated annealing, new search technique, 54–56

single-play games, Ultimatum, 84

software companies, A/B test in, 134–35

Soviet Mathematics (journal), 237n34

Soviet Union, 49

sports, phenomenon of "analytics" in, 4–5

spreadsheet metaphor: linear predictors, 157–58; pattern recognition, 154–55

Stag Hunt, 2x2 tables, 92

standard cells, integrated circuits, 54

Stanford Research Institute (SRI), 165

Stanford University, self-driving car project, 195

statistical analyses, optimal decisions, 6–7

statistical counting, 111, 113

statistical lobbying, Francis's, 117

statistical mechanics, 151

statistical pattern recognition, 145

statistical prediction: actuarial method, 185–86; machine learning, 8; pillar of, 17

statistical rules, dynamic world, 212

statistical significance, 109

statistical tests, 1; personal risk factors, 7

statisticians, on uncertainty, 12–13

statistics: confidence interval, 239n12; heuristic reasoning, 247n47; mammograms, 240n29; null hypothesis, 108–9; standard models, 154; state bureaucrats, 12
Stein, Gunter, on designing control systems, 58–60
steroids, rheumatoid arthritis, 104
Stiebeling, Hazel: on "minimum cost" of diet, 31–32; on affordable diets, 28
Stigler, George: on buying goods in bulk, 232n10; compiling nutrient content, 29; composition of diet, 30–31; Dantzig applying simplex method to diet problem of, 29–30; diet problem as first linear program, 32; optimal diet, 211; on USDA nutrition recommendations, 28
stochastic gradient descent, Shannon's language models with algorithm, 144
stochastic process: sequence of symbols, 149–50; Wiener's theory of, 15
stochastic programming, Dantzig pioneering, 47
strategy, concept of, 65
Strax, Philip: breast cancer death of wife, 121–22; *Early Detection*, 124; leading HIP Study, 122–24
Streetlights and Shadows (Klein), 192, 211
streptomycin: discovery of, 106; outcome of trial, 115; randomized clinical trial (RCTs), 106–10; trial odds, 123
subways, autonomous, 197–98
sulfonamides, 105
Sunstein, Cass, on mathematically rational decisions, 213

SUNY Stony Brook, 165
supercomputer speed, 51
superhuman computer solvers, 98
Supreme Court, Bush and Gore election, 219
Swiss Medical Board, controversial report on mammography, 129–30
symptomatic illness, 115
systematic review, HIP Study, 126–29
system design: creativity, 35; optimization axioms, 34
Szepesvári, Csaba, invention of Monte Carlo tree search (MCTS), 89

TD-gammon, Tesauro's backgammon game, 88
TDIDT (top-down induction of decision trees), 168
technocrats: bureaucracy, 212–13; policy and, 209–10
technology, decision making, 224
temporal difference, value of moves, 77
temporal differencing, 77, 235n11
Tesauro, Gerald, backgammon program TD-gammon, 88
Tesla, Inc., Musk of, 195
Thaler, Richard, on mathematically rational decisions, 213
Theory of Games and Economic Behavior (von Neumann and Morgenstern), 64–65, 78, 83
theory of games of timing, duel theory, 81
therapeutics, innovation in medicine, 104–5
thermostat, control system, 40
Thrun, Sebastien, self-driving car project, 195

TikTok, 140
TimberWolf, 55
TIMIT dataset, (Texas Instruments and MIT), 169
Tomorrow (documentary), 62
truancy, intervention, 137
tuberculosis (TB), 123; outcomes and treatment, 113; streptomycin for, 106–8, 110
Tucker, Albert: game theory group, 79; Prisoner's Dilemma, 83
Tversky, Amos: human decision making, 188; Kahneman and, 188–90, 202
two-player games, 66; "Rock, Paper, Scissors," 66, 67
2001: A Space Odyssey (film), 39

UC Berkeley, 54
UC Irvine Machine Learning Repository, 168
Ultimatum: research on game of, 84; 2x2 tables, 92
uncertainty: quantification of, 16, 17; statisticians on, 12–13
Unisys, 164
United States Preventive Services Task Force, 129
University of Alberta, 95
University of Chicago, 32
University of Illinois, 74
University of Pennsylvania, 75; ENIAC computer, 232n15
University of Toronto, machine learning by team at, 173–74
University of Wisconsin, 165
University of Wisconsin at Madison, workshop, 33, 35
Urmson, Chris, self-driving technology, 195

US Census Bureau, 170
USDA, 32; nutrition recommendations, 28
US Defense Advanced Research Projects Agency (DARPA), 194–95
utility, concept of, 65
Utopia of Rules, The (Graeber), 214

vaccine trials, 114–18; Hill on, 114–16; polio, 116–18
video game, first, playing checkers, 38
Vivalt, Eva, on development economics, 137
von Neumann, John: on chess, 90; formulation of games, 214–15; mathematical activity of, 15; on simple model application, 33–34; "stored program" architecture, 36; theory of games, 78; on theory of games and economics, 64–72
von Stengel, Bernhard, on poker, 94

Waksman, Selman, streptomycin discovery, 106
war, mathematical insight of, 16
war games, RAND Corporation, 39
Waymo: Alphabet on, 250n20; delivering rides, 196
Wayne, Charles, DARPA (Defense Advanced Research Projects Agency) program, 168–69
Weizenbaum, Joseph: on harnessing mathematical rationality, 223–24; on decisions and choices, 221–24, 225
Western Sydney University, 170
whooping cough, vaccine study, 115–16

Wiener, Norbert: mathematical activity of, 15; modeling randomness, 152; signal coding and prediction, 151–52
winning, definition, 2
Women's Health Initiative Trial, estrogen therapy, 132
WordNet, 171
Works Progress Administration, 30
World Series of Poker, 97

World War II, 78, 114, 144; technology, 63

YACR (Yet Another Channel Router), 55

zero-sum games: comparing two-player games, 70; idea of, 66
Zinkevich, Martin, inventing counterfactual regret minimization (CFR), 97